Arcticness

Arcticness

Power and Voice from the North

Edited by

Ilan Kelman

First published in 2017 by
UCL Press
University College London
Gower Street
London WC1E 6BT

Available to download free: www.ucl.ac.uk/ucl-press

A CIP catalogue record for this book is available from The British Library.

ISBN: 978–1–787350–15–1 (Hbk.)
ISBN: 978–1–787350–14–4 (Pbk.)
ISBN: 978–1–787350–13–7 (PDF)
ISBN: 978–1–787350–12–0 (epub)
ISBN: 978–1–787350–10–6 (mobi)
ISBN: 978–1–787350–11–3 (html)
DOI: https://doi.org/10.14324/111.9781787350137

Preface – 'Arcticness and change'

Ingrid A. Medby, UCL Department of Geography

Like a teacher's red pen, the jagged line underneath my writing gave me an uneasy feeling. I tried to ignore it, but the overly conscientious primary school pupil in me would not let it rest: The word processor indicates a spelling error; it's unacceptable to continue, my own internal voice nagged.

'Arcticness' *is* a term, though; and a highly useful one – as I told my word processor with the click of the mouse, 'Add to Dictionary'. Adding the suffix '*-ness*' denotes a state or quality – in this case, the quality of being Arctic.

For those of us interested in the Arctic and, in particular, how people relate to it, a word for the 'quality of being Arctic' is a potential cause for agonisingly many jagged red lines. Although my software clearly disagreed, I am, of course, not the first to see the need for it – something to which this book bears testament. As the world is increasingly looking northwards to a region undergoing rapid change, identifying *what*, *who* or *where* has the 'quality of being Arctic' is high on the agenda; for actors from both near and far, their 'Arcticness' becomes a potential asset as they position themselves for Arctic futures.

But what does it really mean, 'Arcticness'; what are Arctic 'qualities'? Unlike placating a spell-checker, defining what 'is' Arctic (or feeling, believing, thinking, imagining that someone/something/somewhere is) is not as straightforward as it may seem. The region itself is defined in numerous ways depending on topic, context and even interest; and so, determining the *qualities* of a region that cannot itself be fully determined provides a challenge.

Given that claiming an Arctic identity may serve an instrumental purpose – for example adding to political actors' or private stakeholders' credibility – the ambiguity of Arcticness is also in part why the concept is so fascinating, not to mention so *important* to explore. In relations between the Arctic and non-Arctic, the claim to Arcticness potentially becomes a political one; indeed, it may decide who falls on either side of Arctic and the prefixed '*non-*'. In turn, Arcticness becomes a question of who holds rights, who holds responsibilities, and who holds 'true' knowledge of a space in rapid flux ...

Arcticness does not only matter for political decisions and resource extraction; it seems to have become exotic, interesting – it *sells*. With northern lights tours and midnight sun cruises, Arcticness is increasingly commodified. With 'Arctic' labels on anything from bottled drinks to cleaning companies, it has become a brand so ubiquitous that it is now simply part of the everyday.

This has not always been the case. Having grown up in Northern Norway, the change is clear – not *just* climatic or economic change in the region, but a change of label. What was only a decade ago Northern Norway is now frequently referred to as 'the High North' ['*nordområdene*' in Norwegian, translating literally as 'the northern areas'] or the Arctic. A northern identity may now be an Arctic identity – just like our tap water is now 'Arctic water'.

Rebranding the north as 'Arctic' is not for those *in* the Arctic, however, but rather for the outside spectator – tourist, visitor, investor or politician. Speaking of what is Arctic or not, who is Arctic or not, is hardly consequential when you are *there* – it is simply less relevant, less interesting, less exotic. Nevertheless, it is primarily northern communities who face the challenging consequences as the 'frozen' Arctic thaws. What is important to remember here is that these communities have never themselves been 'frozen' (in time), but have always been evolving, moving, changing. Saying that voices from the Arctic are important is not enough – they must also be *listened* to, and finally, engaged in conversation. That is, voices (and ears!) from north and south, east and west, are *all* important in this process of change. Just like a 'new' label, an identification as or with something comes about through negation; and so, Arcticness too takes on meaning through relations and encounters with the constitutive other.

Perhaps then this is more than simply an exercise of marketing or rhetoric; perhaps our concept of 'Arcticness' itself is undergoing change? Could it be that a region which has historically been seen as far away – peripheral to the centre of society – is now being 'drawn closer' due to its accelerating importance to questions of climate change and globalisation?

Interrogating *why* something is now considered 'Arctic' is highly important; as is attention to *who* claims so, or who is now themselves considered Arctic: these are questions of power – as noted, power both to speak and to act. But more than this, Arcticness under change may point to a more profound change in our relations to a region, to ourselves and to each other. It may be symptomatic of ever more people feeling that the Arctic *matters* – also to those living far south of the Arctic Circle. As the adage goes, what happens in the Arctic does not stay in the Arctic; and vice versa, Arctic change does not have its origins in the Arctic either. In other words, it points to a realisation of our interconnectedness – one that has always been there, of course, but which is now far more visible and *felt* thanks to satellites, the internet, travelling and globalisation writ large.

In the end, Arcticness *cannot* be easily defined; no more than the region itself can be neatly placed within latitudinal lines. It is as much about relationality – both at the level of diplomatic negotiations and that of daily life. And indeed, Arcticness perhaps *should not* be limited to semantic boundaries, should not be rendered static on the pages of a dictionary. Rather, it should be kept open – open to interpretation by those to whom it feels relevant.

The above 'challenge' of determining Arctic qualities is also an opportunity: An opportunity to think beyond boundaries – or *without* them altogether; to think and imagine anew for alternative ways of understanding. The Arctic is, as the following chapters will discuss, undergoing profound change due to climate change, globalisation and many other influences – and so is, and should, our concept of Arcticness. It is through interaction, through relating to each other, that the '*ness*' – the quality of *being* anything at all – takes on meaning.

List of figures

Rachel L. Tilling works as a Research Fellow at the University of Leeds, having completed her PhD in Polar Remote Sensing at UCL. Rachel's research focuses on satellite observations of Arctic sea ice. These are combined with measurements collected as she has camped, flown and sailed across the frozen Arctic Ocean.

James Van Alstine is Associate Professor in Environmental Policy and Co-Director of the Sustainability Research Institute at the University of Leeds. With a disciplinary background in human geography, James' research focuses on natural resource governance and politics and climate change governance in the global North and South.

Vladimir Vasiliev was born in a very small village, Tuora-Kel, in Central Yakutia. After university, he worked as a teacher in a secondary school, then at the Institute of Biological Problems of the Cryolithozone of the Siberian Division of the Russian Academy of Sciences. For more than 10 years, he served as the Deputy Director in the Yakut International Centre for the Northern Territories Development and worked in the Ministry on Nature Protection. Since 2000, he has been involved closely in Northern Forum activities and worked as the Northern Forum's Executive Director in 2012–2014. He was invited to be the Minister on Federal and External Relations, Sakha Republic (Yakutia), Russia in October 2014.

Emma Wilson (PhD, Cantab, 2002) is an independent researcher/consultant, director of ECW Energy Ltd., and Associate of the Scott Polar Research Institute, University of Cambridge. She has 20 years' experience in extractive industries and community relations, including social impact assessment and anthropological fieldwork in Russia, Uzbekistan, Norway, Greenland and Nigeria, among others.

Tun Jan (TJ) Young is a PhD Student in Polar Studies at the University of Cambridge. His research integrates electrical engineering, numerical modelling and field glaciology to investigate the basal and englacial regimes of the Greenland Ice Sheet.

1
Editorial Introduction: Shall I compare thee to an Arctic day (or night)?

Ilan Kelman

Arcticness as a home

People and communities, lives and livelihoods. These define the Arctic, just as with all other populated areas on the planet. Is there, then, anything special, specific, exceptional or unique about the Arctic? To the peoples in the Arctic, the answer is 'of course'.

Because it is home.

As Arctic literature is fond of stating, there is no single Arctic. Definitions abound, from being a region or place to being an idea or phenomenon. The Arctic is delineated by latitude, tree lines, national and subnational borders and indigenous territories, among many other suggestions. All these elements vaguely concentrate into the northern areas of Canada, Finland, Norway, Russia and Sweden along with all of Alaska, Greenland and Iceland.

This is the Arctic as a place – and the Arctic as place. The Arctic is also characterised, perhaps more so, by its people. Depending on where boundaries are set exactly, the Arctic's population is anywhere from approximately 4 million to approximately 13 million people. About 10 per cent of Arctic inhabitants are indigenous, belonging to 40 different groups, examples of which are Saami, Inuit, Nenets, Yakuts and Aleuts. In some jurisdictions, such as Nunavut and Greenland, indigenous peoples are the majority. All Arctic areas have comparatively low population density.

Arctic indigenous peoples are partly defined by the way in which they were colonised from the south. Iceland is the only Arctic country without designated indigenous peoples. The other seven countries have never fully addressed their post-colonial legacy which included active suppression of indigenous languages and cultures, forcing nomadic peoples to settle, and taking indigenous children away from their families for the purpose of 'education' and 'acculturation'.

As part of aiming to re-connect Arctic peoples and places, and to redress past mistakes, each post-colonial Arctic country apart from Russia has, to a large degree, settled land claims with Arctic indigenous peoples. The settlements occurred in different ways and in different time periods, with implementation, monitoring and enforcement still not fully functional in many instances.

The generational context adds complexity. The generation of leaders who grew up under colonialism and who negotiated the settlements are now in the process of retiring. They are giving way to a new generation of leaders who did not experience similar difficulties or frontline fights for autonomy and the recognition of indigenous cultures. They face other challenges, such as low educational attainment, high rates of substance use and abuse, and high suicide rates.

They are also looking to connect to the world beyond their (mis) governing state through the internet and social media to define and re-define, and to be proud of, their indigeneity, their peoples and their places; that is, their Arctic. The battles are not over. Greenland's independence is still a possibility. Racism against indigenous peoples remains. The peoples are not homogeneous groups, such as the Saami who have different livelihoods including reindeer herding, fishing, both and neither.

Non-indigenous Arctic peoples also represent the Arctic, not just Icelanders but also those born and/or living in the north but without an Arctic indigenous heritage. One class of Arctic peoples, most notably in Scandinavia, comprises immigrants from around the world, including refugees, who fully settled in the Arctic and who are now raising first-generation, Arctic-born families with diverse, international heritages.

Within this Arctic rainbow, what is the Arctic? How do Arctic peoples relate to their places? The ways include living, livelihoods, environments and movements. In many locales, movement means the typical commute by private or public transport to a nine-to-five office job. In many locales, it is the typical subsistence hunting, conversing with the wind, feeling the sea, traipsing the land and traversing the ice.

Water (solid and liquid) and wind flow, bringing with them life and death. The Arctic peoples flow with them. Movement, survival and thriving are choreographed within the elements and within the colours of the seasons: blue, grey and white melding with brown, green and splashes of colour in summer flora and fauna. The ever-changing kaleidoscope of weather and skies, of animals and oceans, of plants and the Earth, creates Arctic flows and ebbs.

Transitions and boundaries are prominent but fuzzy. Snow melds into land shifts to water becomes ice, drifting lazily under the dazzling dome of the summer sun and the scintillating stars of the wild winter. When the ice roads thaw making transport difficult, inland communities are spoken of as being landlocked. When the ocean is too rough for boats and the wind is too dangerous for planes, island communities are seen as being entrapped.

What vocabulary suggests being icelocked? The ice can be too thin on the water or too crevassed on the land, or just too slushy everywhere. The transition between seasons can be harsh when the land ice and sea ice mixtures do not permit safe transport. Then, one's Arctic place becomes evident, as an islander or not, as someone who enjoys being indoors or not.

Movement and entrapment mean that Arctic placeness is not contentedly fixed. In any case, the glaciers, the ice, the snow, the water and the wind are always in motion. The rivers and the seas emote ripples and waves. The tides breathe for the water and the wind for the air. Coasts erode and accrete – with both ice and sediment.

Arctic changes are expressed in other ways. From colonisation to self-determination, the Saami have created their parliaments, referenda supported autonomy for Greenland and Nunavut, and Russian regions and territories have various levels of self-governance. Exceptionalism identifies many Arctic place traits – including the internationally unique Svalbard Treaty and the central Bering Sea having its 'donut hole' which is an enclosed polygon of international waters surrounded by territorial seas.

The scale of Arctic territories is sometimes forgotten. From Murmansk to Chukotka, the time difference across Russia is nine hours. Alaska has only two time zones, an artificial construction, but as the largest American state more than twice the area of its nearest rival, it is almost as wide and as tall as the entire contiguous states. Ottawa–Iqaluit flights travel more than three times as far as the London–Edinburgh route and are still shorter than Greenland's full north–south distance.

Current national borders across the Arctic are poorly reflective of indigenous cultures. The Saami are partitioned among four countries.

Only modern politics draw a line between Alaska and Yukon. The Canada–Denmark dispute over Hans Island is meaningless for peoples who use the land, sea, ice and wind to live.

Many of these Arctic placeness discussions are characterised by islands and archipelagos including the Aleutians, Hans Island, Greenland, Iceland and Svalbard. Nunavut's capital sits on Baffin Island rather than the mainland. Many of Norway's principal Arctic settlements are on islands including Tromsø, Harstad and Hammerfest.

Island studies has evolved over the past generation, exploring the natures and personalities of islands, island communities and islanders. Much debate and critique has centred around what it means to be an island or an islander, defining and examining the essence of islandness. These and similar questions and explorations have emerged for the Arctic, Arctic communities and Arctic peoples.

Thus, we generate and query the term Arcticness through the chapters in this book.

Arcticness as a book

The chapters here birth, live and quash Arcticness in differing tones and styles. Disciplinary and non-disciplinary examinations range from geophysics to law, from anthropology to engineering and from art to resource management. Personal experiences and internal realities sit alongside technological investigations and external observations and representations. The transitions among the chapters can be as jarring as Arctic weather changes, as mismatched as some northern and southern views and as manifold as the Arctic landscape.

The Arctic breathes diversity and Arcticness embodies variety. The chapters in this book reflect this range through poetic interludes alongside detailed social and physical science interspersed with images confiding more than a thousand words meshed with lengthy policy grillings. Some chapters dive deeply, unearthing (or deicing?) the authors' tiny yet vast Arctic worlds. Others prefer breadth, traversing continents and disciplines to comparatively analyse locations within and outside of the Arctic of Many Definitions.

Consequently, the chapters Arctic-hop – around, through and within longitudes, latitudes, ideas, modes, genres and especially peoples. The Preface and Afterword frame this collection through personal reflections of being Arctic.

As an ensemble, these contributions – but, more so, the peoples penning them – probe Arcticness, including technical and place-based standpoints, involving northern and non-northern viewpoints (and their combinations), and incorporating science, policy and practice – but all with the fundament of the human perspective. Because Arcticness and all the chapters herein are still a human construct, emerging from and being forced on people, and occurring within a human context.

Arcticness as a context

Phrases other than Arcticness are feasible. The term 'islanders' questions why 'Arcticers' does not exist, instead referring to Arctic peoples along the same lines as island peoples. Arctic provides both a noun and an adjective, with other terms such as Arcticesque and Arcticite not being considered, appearing both awkward and vapid, even platitudinous, trying to construct something Arctic which mirrors little. Translation difficulties, particularly into Arctic languages, would also result from these artificial constructions.

Yet artificiality itself is not necessarily disingenuous or disadvantageous. Humans have a right and a need to create ideas regarding their places, their movements, their livelihoods, their peoples, their environments and their homes. The challenge and opportunity, as with Arcticness, is whether or not the artificial creation has real and useful meaning.

We should not Arcticise for the sake of finding, generating or discussing Arcticness. Where potential exists for substantive idea and action, it deserves examination. This is the case with Arcticness.

Arctic imaginaries, Arctic realities and their intersections in and outside of the Arctic pervade numerous historical, contemporary and future discussions. From the establishment of Arctic peoples to exploration and colonisation from the south to re-establishing sovereignties and Arctic peoples' control over themselves, Arcticness displays tangibility and ephemerality. Meanwhile, non-Arctic peoples try to wrest control and make Arcticness relevant for themselves, from the construction of 'last-chance' tourism to tropical countries seeking observer status at the Arctic Council.

The authors in this book recognise this gamut. They accept what they understand and do not understand, what they have and have not experienced. They have reached into their science and reached into their soul, writing from the head and writing from the heart. Their chapters show how Arcticness portrays and betrays the Arctic, its places, its peoples and its homes. Even when they do not come from the north, the authors seek its power and voice – to understand, learn about, explore, compare, apply and critique Arcticness.

PART 1
Arcticness Emerging

2
Maintaining my Arcticness

Heather Sauyaq Jean Gordon

My name is Heather Sauyaq Jean Gordon. My Iñupiaq name, Sauyaq, means drum in Alaska Seward Peninsula Iñupiaq. I was given this name as an adult, when I was 31 in 2016. My Great Aunt Peggy Perry (Aunt Peg) died, and my Grandmother passed her name on to me. I work daily to speak out, be heard and carry the beat of my culture in my heart. My Aunt Peg was an opinionated and hilarious woman. I strive to keep her alive in me.

Aunt Peg was a beader. She beaded for around 15 to 20 years before she died. She beaded earrings, necklaces and bracelets, as well as a few other ventures that she tried out when interested. I always loved her beading. When I was in my late teens, she taught me to bead a strap to hold glasses around my neck. I never finished that project. However, I always wanted to be able to bead like her.

In 2015, I was six years into graduate school and struggling with knowing who I was, and who I wanted to be. I felt detached from the original reason I went to graduate school, to become a professor. I wondered if that truly was a way to best help the Iñupiaq people. I was feeling that being in school was potentially a selfish endeavour and was concerned about what route I should take. In the fall of 2015, I came across the Indigenous Studies programme at the University of Alaska Fairbanks. This programme reinvigorated my interest in self-determination, sustainability and well-being in the lives of Iñupiaq people. Yet, I still felt something missing. I now knew what I wanted to be, an advocate for self-determination and well-being. I still did not know who I was.

I knew I was an Iñupiaq woman of mixed heritage. I knew I grew up in Homer, Alaska, outside of Iñupiaq lands. Yet, I did not feel separated from the land, as I grew up raising reindeer and living a subsistence lifestyle. I travelled to Alaska Native villages to do construction work while I was in high school. In the villages I was exposed to Alaska Native beading. It was beautiful.

I conducted my master's research in Greenland from 2010 to 2012. While there, I met fantastic artists. The women beaded beautiful National costumes, earrings and even coasters for coffee mugs on the table. Their art was inspirational. These women were my Inuit cousins who had travelled from Alaska many years earlier to settle in Greenland. I felt a connection with them and their art. I also was able to eat Native food in Greenland: seal, whale, fish and much more. These experiences brought back my tie to the Arctic and the Arcticness I shared with Indigenous people in the circumpolar North, made me miss Alaska, and made me question what it meant to me to be an Inuit woman.

Now, in 2015 I was living in Madison, Wisconsin, as I had been since 2009 for my master's programme. I knew few Native people and no other Alaska Natives in the area. Facebook seemed like my main connection to Alaska, except for the trips home I would make in the summer for fishing and berry picking. I sought help from a Ho-Chunk trained non-Native counsellor and she taught me about the four aspects of a healthy life: spiritual, mental, physical and emotional. My life met the mental, physical and emotional but not the spiritual aspects. I had little connection to my Iñupiaq culture and felt disconnected from Alaska while living in Wisconsin.

I looked back at my knowledge of Iñupiaq culture. I thought of the beading I learned from my Aunt Peg, the women in the rural villages I visited while doing construction and the women in Greenland. I felt that beading was my connection to Alaska and being Iñupiaq. It was my connection not only to my Indigenous identity but also to my sense of Arcticness while living so far from the North. So, I began beading earrings.

In the spring of 2016, my Aunt Peg died. It was then that I was given her name. I felt her spirit within me and wanted to continue on her legacy of beading. When we were going through her home, I was given the opportunity to go through her beading room. Her beads were everywhere. There were patterns, needles and just everything. I felt her presence and it made me so happy that I could continue doing for her what she loved. I chose beads that would allow me to do work similar to what she had been doing. It was then that I came across a beaded glass ornament. It was absolutely stunning. When I came home I started beading ornaments in addition to the earrings I already made. I want to keep her memory alive.

I had begun beading in the fall of 2015. After my Aunt died, beading became very important to me and is now a part of my life that I regularly practise (Figures 2.1 and 2.2). It calms me, makes me feel the

Fig. 2.1 Beaded forget-me-not earrings (Source: author).

Fig. 2.2 Beaded glass ornaments (Source: author).

Arcticness I have, even though I live far from Alaska, and makes me feel connected to my Aunt Peg. I have started selling the art, like she did. It makes me happy to see my spiritual connection to the Arctic make others excited and happy. I feel they are spreading Arcticness to each person they meet as they share my beading.

I enjoy giving away my art to family and friends. I feel a connection with each person who wears my art. The time I put into the piece, the spiritual connection to being Iñupiaq, and working in the memory of my Aunt makes every piece I produce special and unique. My beading spreads the beauty of Arctic lifeways and artistic patterns across the world.

3

Conversations in the Dark

Larissa Diakiw
(publishing as Frankie No One)

Conversations in the Dark is the first in a series of graphic essays that follows Frankie as she explores history. In this comic she reads the food chapter in Canada's Truth and Reconciliation Commission, a report that details the abuses that took place under the residential school programme.

IT IS SPRINGTIME. I AM SITTING OUTSIDE A FRIEND'S FARM, UNDER A FINGERNAIL MOON. MAI READS ME EXCERPTS FROM THE TRC*.
* THE TRUTH & RECONCILIATION COMMISSION IS AN INQUIRY INTO THE LEGACY OF RESIDENTIAL SCHOOLS IN CANADA. ESTABLISHED IN 2008 & COMPLETED IN THE SUMMER OF 2015, IT REVIEWS THE ATROCITIES COMMITTED BY THESE SCHOOLS & THE CANADIAN GOVERNMENT, COMPILING THE TESTIMONIES OF OVER 6000 SURVIVORS IN ORDER TO START A DISCUSSION ON WHAT RECONCILIATION & DECOLONIZATION MIGHT MEAN...

THE TRC EXAMINES THIS BRUTAL PROGRAM OF CULTURAL GENOCIDE.
CAN I READ PART OF THIS TO YOU?

IS IT SAD?
WELL, YEAH BUT IT'S IMPORTANT TO HEAR.
WE GREW UP IN ALBERTA. MAI'S MOTHER CAME TO THE PRAIRIES ON AN AIRPLANE FROM THE PHILIPPINES.
MY GRANDPARENTS FLED THE OVERCROWDED FARMLANDS OF EASTERN EUROPE BY BOAT.
UNDER THE CHIME OF INSECTS WE CAN HEAR THE ABSENCE OF BISON, WHICH WERE KILLED OFF ALMOST ENTIRELY AFTER CONTACT.
SOME ESTIMATE THE BISON POPULATION WAS AROUND 60 MILLION, & BY 1890 ONLY 540 WERE LEFT SCATTERED IN THE WOODS & HUDDLED TOGETHER IN A YELLOWSTONE CAVE.

GOV'T POLICIES ENCOURAGED THE STRATEGIC SLAUGHTER OF THE BEASTS. BOUNTY HUNTERS STALKED THE GRASSLANDS, SHOOTING FROM TRAINS, WAGONS, HOPING FOR DOLLARS IN EXCHANGE FOR HIDES. POWERED BY THE OPIATE OF MANIFEST DESTINY, & ULTIMATELY MAKING WAY FOR A NEW AGRARIAN CAPITALISM, THEY CULLED THE ANIMALS WITH FRIGHTENING PRECISION.

THEY KNEW THAT THIS
WOULD LEAD TO
STARVATION FOR PEOPLE
WHOSE LIVES RELIED
ON THE MEAT.

AT THE SAME TIME THE GOVERNMENT DEVELOPED A
NETWORK OF BOARDING SCHOOLS FOR INDIGENOUS
CHILDREN, INFAMOUS FOR THE HORRORS THAT TOOK PLACE
BEHIND THEIR DOORS.

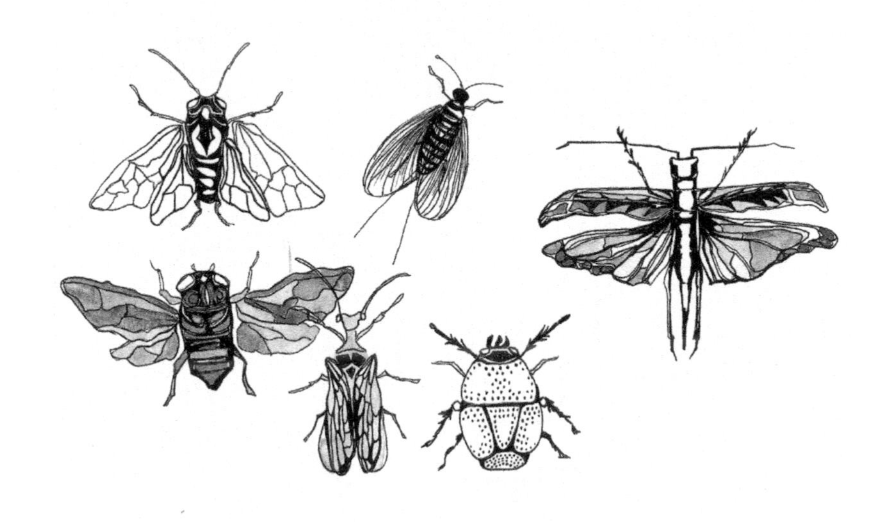

HUNGER
WE "WERE ALWAYS HUNGRY."
- ENOS MONTOUR,
SURVIVOR OF MOUNT ELGIN
RESIDENTIAL SCHOOL
MUNCEY, ONTARIO

"THE FOOD DIDN'T TASTE VERY GOOD BECAUSE WE DIDN'T HAVE OUR TRADITIONAL FOOD THERE, OUR MOOSE MEAT, OUR BANNOCK, OUR BERRIES."
- DAISY DIAMOND, SURVIVOR OF SHINWAUK RESIDENTIAL SCHOOL, SAULT STE. MARIE, ONTARIO.
"PRIESTS ATE THE APPLES, WE ATE THE PEELINGS, THAT IS WHAT THEY FED US WE NEVER ATE BREAD."
- GLADYS PRINE, SURVIVOR OF SANDY BAY RESIDENTIAL SCHOOL, SANDY BAY FIRST NATION, MANITOBA.

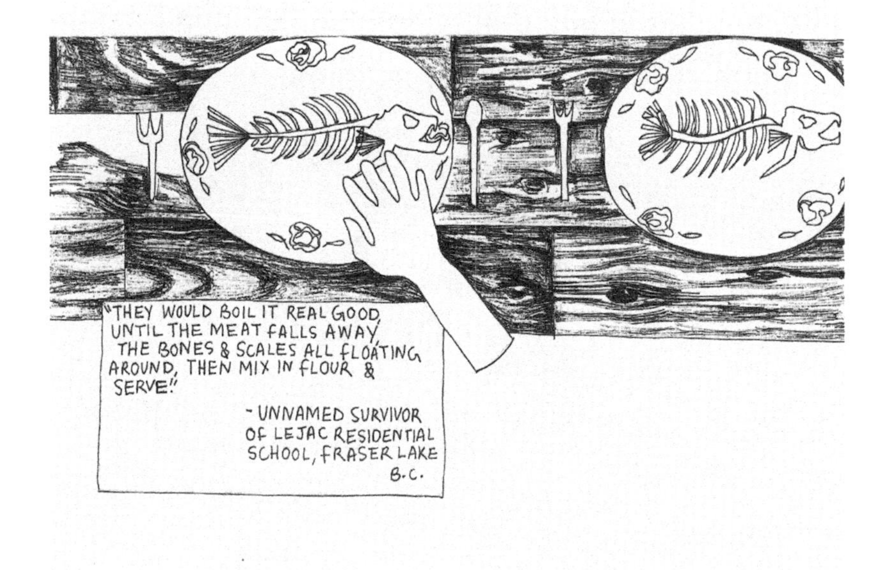
"THEY WOULD BOIL IT REAL GOOD, UNTIL THE MEAT FALLS AWAY, THE BONES & SCALES ALL FLOATING AROUND, THEN MIX IN FLOUR & SERVE."
- UNNAMED SURVIVOR OF LEJAC RESIDENTIAL SCHOOL, FRASER LAKE B.C.

"THERE WAS A LOT OF TIMES I SEEN OTHER STUDENTS THAT THREW UP & THEY WERE FORCED TO EAT THEIR OWN, THEIR OWN VOMIT."
-BERNARD CATCHEWAY, SURVIVOR OF PINE CREEK RESIDENTIAL SCHOOL, PINE CREEK MANITOBA

WHO COULD FIGHT AN INVADING GOVERNMENT WHO HELD THEIR CHILDREN HOSTAGE?

WHO CAN FIGHT WHEN THEY ARE ALWAYS HUNGRY?

I USED TO BABYSIT THIS CREE KID WHO WAS IN FOSTER-CARE. HE WAS TINY & HAD THESE GIANT BOTTLE-CAP GLASSES. EDWARD.
WE WATCHED MOVIES TOGETHER ON SATURDAYS, LIKE GHOSTBUSTERS. THE POOR KID HAD BEEN NEGLECTED FOR LONG PERIODS OF TIME...

& HE WAS AFRAID OF STARVING AGAIN, SO HE HID FOOD EVERYWHERE: UNDER THE BED, IN JACKET POCKETS, IN CUPBOARDS, BEHIND THE T.V.
DID IT ROT?
YEAH, SOMETIMES YOU COULD SMELL IT.

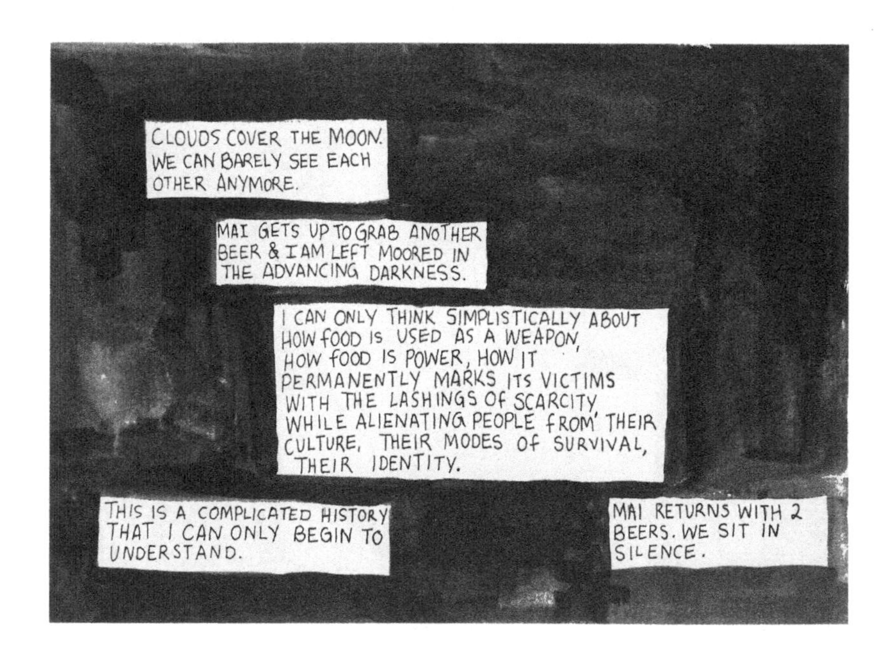
CLOUDS COVER THE MOON. WE CAN BARELY SEE EACH OTHER ANYMORE.
MAI GETS UP TO GRAB ANOTHER BEER & I AM LEFT MOORED IN THE ADVANCING DARKNESS.
I CAN ONLY THINK SIMPLISTICALLY ABOUT HOW FOOD IS USED AS A WEAPON, HOW FOOD IS POWER, HOW IT PERMANENTLY MARKS ITS VICTIMS WITH THE LASHINGS OF SCARCITY WHILE ALIENATING PEOPLE FROM THEIR CULTURE, THEIR MODES OF SURVIVAL, THEIR IDENTITY.
THIS IS A COMPLICATED HISTORY THAT I CAN ONLY BEGIN TO UNDERSTAND.
MAI RETURNS WITH 2 BEERS. WE SIT IN SILENCE.

Tracking the Arctic

Wrenched white vista winds
Knead them black, turn it around
Tracks were stretching south

Funsho Martin Parrott

Tracking the Arctic

Funsho Martin Parrott

Background

Haiku is a Japanese poem of 17 syllables, in 3 lines of 5, 7 and 5. Haiku traditionally evokes images of the natural world. The purpose of Haiku is to create a poetic form which has the brevity and intensity of the moment. The rigour of Haiku ensures that only the most essential parts of an idea or moment remain.

Rationale

The theme of this Haiku is irreparable change and crossing a point of no return. The first line speaks about the damaged landscape, with the repetition of 'W' representing the ice caps and glaciers of northern terrains. The second line seeks to illustrate the unforgiving nature of the Arctic climate with the 'Knead'. 'Knead' is a *double entendre*. If read as 'need', the line becomes about our insatiable need for 'black' or oil, while 'knead' describes the forceful folding and distorting one can experience within Arctic winds. The last line also has a double meaning. It speaks about the footprints that Arctic wildlife would have left in the ice and snow, had it not been for the damage to their habitats; hence, their description in the past tense. It also references the work of Arctic organisations and places it alongside the art of tracking, a skill belonging to many First Nations people. They track within the Arctic; we track the Arctic itself.

4
Radar observations of Arctic ice

Rachel L. Tilling, Tun Jan Young, Poul Christoffersen, Lai Bun Lok, Paul V. Brennan and Keith W. Nicholls

Background

To many observers, the Arctic is synonymous with snow and ice. For example, the Arctic Ocean spans just over 14 million km^2 – an area larger than that of Europe – which is variably covered in frozen ocean water, or sea ice, throughout the year.[1] The Arctic Ocean is almost completely surrounded by land, which can often be covered in snow, permafrost (frozen soil, rock or sediment) or land ice. Observing how different forms of ice in the Arctic are changing, and understanding how they have evolved in the past, is crucial. Radar technology provides us with a tool to do this and allows us to visualise the glacial environment beneath the ice surface. This chapter provides an overview of modern radar-based observation methods and describes how measurements from them have contributed to a scientific understanding of ice with an emphasis on Arcticness.

Over the past few decades, radar technology has significantly contributed to our understanding of the Arctic landscape. It has become a key tool in observing changes in the Arctic snow and ice cover. For example, data from radar satellites have been used to document changes in the thickness of the sea ice that covers the Arctic Ocean, by measuring the elevation of the ice and ocean surfaces separately. These data are now available in near real time (NRT) and will allow us to assess Arctic environmental change as it is happening. They also have the potential to help industries such as tourism and transport to navigate the polar oceans with safety and care. Radar-based measurements can be used to constrain the physics of ice flow within models that predict the future state of Arctic ice and global climate.

By using a growing suite of satellite, airborne and ground-based radar data, scientists are able to study the past Arctic climate, provide information on the present state of the Arctic and aid future predictions of Arctic climate change. Communicating such Arctic science – to other scientists, the media and the public – and integrating the physical sciences and engineering with Arctic action on the ground raises further challenges which need to be overcome to deal with Arctic change. As scientists, it is this ever-changing, physical landscape that on first thought embodies the concept of 'Arcticness', and the ability to use scientific tools to observe and quantify these changes. On reflection, this is a rather remote and emotionless connection – like the connection between a scientific instrument and the landscape it observes.

One must consider Arcticness internally and not just externally, to truly feel a connection to it, and realise that the Arctic is so much more than a distant land to be studied from afar. The Arctic is livelihood; it is support despite a lack of physical contact; it is passion and excitement; and it is the draw of the supposed unknown. Never does that become clearer than when standing on the ice and experiencing overwhelming silence interspersed with deafening creaks and groans. No science can describe the feeling of isolation and exhaustion, which is far outweighed by sheer euphoria and awe.

Radar observation methods

Radar – an acronym for Radio Detection and Ranging – functions by transmitting and receiving pulses of radio waves to investigate the location and properties of a target. The distance from the radar instrument to a target can be determined by measuring the round-trip delay of a radar pulse – a simple concept that originated from the classical experiments conducted by James Clerk Maxwell (1865) and Heinrich Hertz (1886), reflecting a 455MHz wave off metallic objects from a distance. However, it is inappropriate to ascribe the *invention* of radar technology to a specific incident or date, but instead through over a century of developments and refinements of radio technology, and aided by key geopolitical events during the mid-twentieth century.[2] Although the concept of radar is overwhelmingly associated with World War II, the prominence of ionospheric research before and after the war, as well as the Space Race behind the scenes of the Cold War cannot be ignored.[3] Initially using radar to measure the height of the ionosphere, this scientific technique was then adapted by Sir Robert Watson-Watts in 1935 to detect aircraft at a distance, along with other radiolocation methods developed by other countries under the cloak of military secrecy. The conclusion of World War II saw the dawn of radar for non-military use, with notable developments in civil aviation, meteorology, astronomy and geology.[4]

One of the earliest radar instruments used in the Polar Regions was ground-based ice penetrating radar, to investigate Antarctic ice shelves in the 1960s.[5] Since then, the interest in radar turned towards airborne and then satellite platforms to give a wider view of the ice. The characteristics of the radar waveform used vary depending on its application. High power short pulsed waveforms are very commonly used while, more recently, low power swept-frequency waveforms (pervasive within the automotive radar community) are popular particularly within research in academia due to their low cost and ease of implementation. Often, the primary goal of many radar surveys on or above ice is to measure ice thickness, which is an important component in models of ice dynamics, ocean circulation, global heat budgets and sea-level rise. However, ground-based radars can also investigate certain qualities within ice, such as changes in crystal structure, internal layering and water content.[6] The following section describes how different radar instruments can be used, with examples relevant in the Arctic. Principally, radar is used in two different ways to study the properties and characteristics of snow and ice. The first is to obtain data on the changes of one area through time. The second is to obtain an 'image' of an area to be considered.

Satellite radar

Radar instruments on board satellites can be used to observe the changing Arctic ice cover, using satellite radar altimetry. Altimetry is a technique in which the height of an instrument is measured above a target surface. In satellite radar altimetry the distance is measured from the satellite to the Earth's surface. The Earth's surface elevation can then be calculated by combining the distance measurement with precise knowledge of the satellite's orbit. In the Arctic, satellite radar altimetry has been used to measure the elevation of land ice,[7] the oceans[8] and sea ice.[9] Applications of the data include estimating the mass of ice sheets,[10] ocean circulation[11] and sea ice thickness and volume.[12]

Sea ice covers about 12 per cent of the world's oceans. Because of its salt content, ocean water begins to freeze when it reaches a temperature around –1.8°C. Sea ice is a major element of the Earth's climate system. It regulates atmospheric temperature by reflecting the sun's energy and by forming an insulating layer between the ocean and the atmosphere. This insulating layer slows heat exchange from the relatively warm ocean to the cool atmosphere. The growth and melt of sea ice also affects freshwater input into the world's oceans. As sea ice grows, salt is expelled in a process known as brine rejection and as sea ice melts, relatively fresh water is released to the ocean. In the northern hemisphere, Arctic sea ice regulates the freshwater input into the Arctic Ocean and the subpolar North Atlantic. The Arctic temperature and freshwater balance affect patterns of atmospheric and oceanic circulation across the region and at lower latitudes. These in turn could impact on the climate in Europe, America and across the northern hemisphere through, for example, changes in rainfall[13] or an increase in extreme weather events such as drought and flooding.[14] To fully understand the global impacts of changes in the Arctic sea ice cover, long-term and accurate observations of the entire ice pack are required. It is now possible to measure the thickness and volume of sea ice across the Arctic by using satellite radar altimetry.

In 2010 the European Space Agency (ESA) launched the CryoSat-2 satellite.[15] CryoSat-2 was, and still is, the only radar

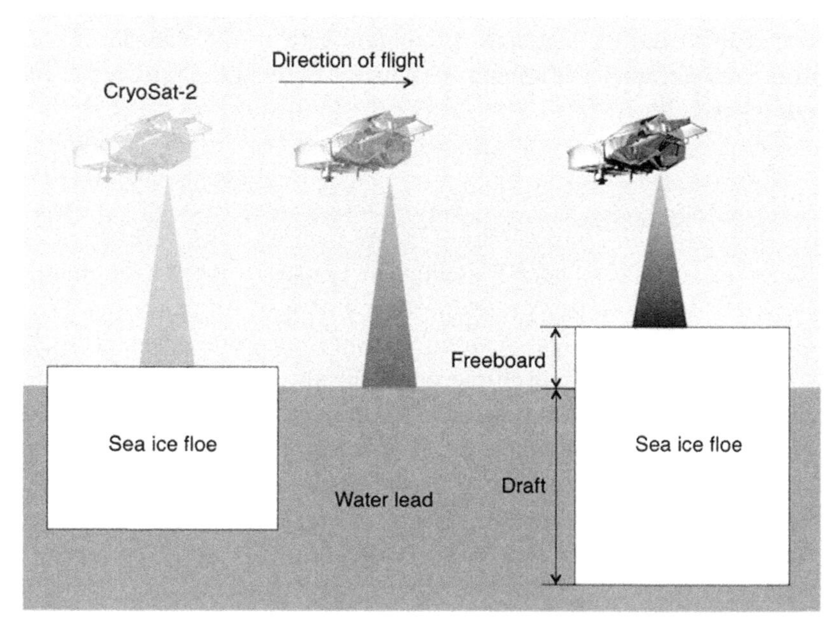

Fig. 4.1 Measuring sea ice thickness from the CryoSat-2 satellite (Source: author).

altimeter satellite to cover close to the entire northern hemisphere. It provides unparalleled coverage of the Arctic Ocean with an orbit that extends to 88°N. The resolution of the CryoSat-2 radar allows it to measure the elevation of Arctic sea ice, and the elevation of the open water in the cracks between sea ice (Figure 4.1). The ice blocks are called *floes* and the cracks are called *leads*. The difference in height between these two surfaces is known as the sea ice *freeboard* – the elevation of the sea ice above the ocean surface. A buoyancy calculation can then be applied to estimate the thickness of the ice below the waterline, which is called the sea ice *draft*. Combining the freeboard and draft gives the total ice thickness. Sea ice volume is simply the ice thickness multiplied by area.

Airborne radar

Compared to satellite radar observations, measurements of Arctic snow and ice conducted through airborne radar missions produce images at a much higher spatial resolution but at the cost of a smaller footprint.[16] This can be beneficial in studies within single glacial catchments, for instance, to investigate local topography and how the shape of the terrain influences the flow and deformation of the ice above. This is a key parameter in models predicting loss of land ice and sea-level rise.

The majority of airborne radar is 'side-looking', whereby antennas are carried underneath the aircraft body or wings and are fixed to look at right angles to the aircraft's trajectory.[17] Continuous scanning while the aircraft is moving produces images that overlap in space, which can then be stitched together to create a composite swathe of the study area in question.

Much side-looking radar takes advantage of the moving platform by using one antenna in time-multiplex, which creates a synthetic antenna aperture consisting of the same antenna receiving echoes in a lengthwise array. This technique allows Synthetic Aperture Radar (SAR) to produce images with extremely high (millimetre) resolution, which differentiates it from other traditional airborne radar.[18] Normally, the output image resolution is scaled with the size of the antenna aperture – this is analogous to using larger telescopes to explore regions of outer space. Therefore, SAR is a convenient way to obtain measurements with a spatial resolution at millimetre accuracy, which would otherwise require an impractically large (greater than 10 m) antenna array.

A noteworthy dataset obtained from airborne radar are the observations from NASA's Operation IceBridge[19] – an eight-year-long mission of the largest airborne survey of the Earth's Polar Regions. The objective of the mission is to bridge the gap in polar observations between the unexpected degradation of ICESat's Geoscience Laser Altimeter System (GLAS) in October 2009 and the planned launch of the replacement ICESat-2 in late-2018.[20] IceBridge is a vital component that ensures a 20-year continuous record with the advent of ICESat in 2003 and the estimated design life of ICESat-2 ending in 2022.[21] IceBridge operates using a number of research aircraft that are equipped with radars, laser altimeters, photographic mapping systems and tools to measure surface gravity and magnetic properties.

Ground-based radar

Radio echo sounding (RES), another form of radar, functions by transmitting and receiving electromagnetic waves at specific frequencies or frequency bands in order to investigate the properties of ice.[22] The propagation of radio waves through ice is principally controlled by the *permittivity* and *conductivity* of the ice material. The contrast between these properties and other materials, such as water and sediments, causes a proportion of the radio waves to be transmitted through the ice to reflect back to the radar receiver on the surface. This delay and a (multiple) reflection mechanism allow the ice base and layers within the ice mass to be observed non-invasively by appropriate processing of the received radar signal. In the late 1990s, interests in ground-based sounders, in particular phase-sensitive radars (also known as pRES), began to re-emerge at the British Antarctic Survey[23] after their initial popularity in the 1960s (see earlier). Such systems have many advantages over mobile platforms. For instance, a stationary system can acquire many radar datasets that can be coherently averaged to improve the signal-to-noise ratio. This will then help to reveal the weak internal reflectors within the ice column. It also makes it possible to use a low transmit power, often three orders of magnitude lower than airborne systems. Another popular application of ground-based radar is to profile ice sheets along transects.[24]

Recent advances have been made in a new phase-sensitive radar based on frequency modulated continuous wave (FMCW) techniques for glaciological applications.[25] A version of this radar has also been configured with an array of antennas to form an experimental system that allows imaging through the ice.[26] A block diagram of a typical ice monitoring radar system is shown in Figure 4.2.

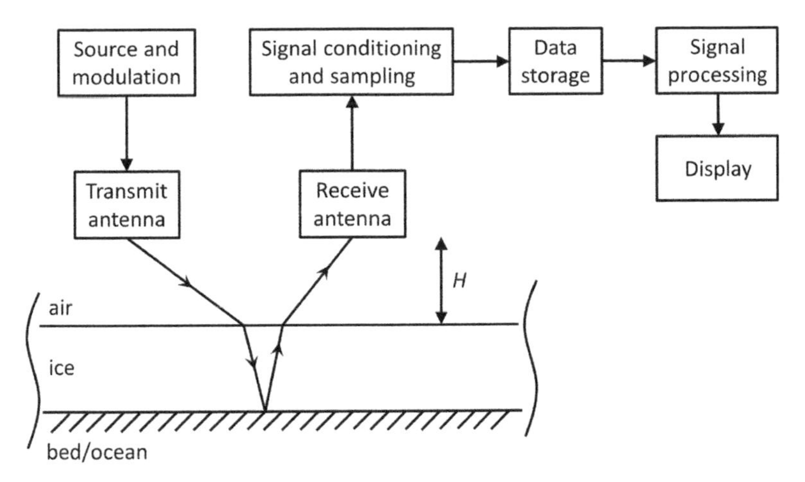

Fig. 4.2 Block diagram showing operation of a typical ice monitoring radar (Source: author).

Scientific examples of Arctic ice observations

Satellite observations of Arctic sea ice thickness

Satellite radar altimetry can be used to measure Arctic sea ice thickness, as discussed in the previous section. When combined with estimates of the sea ice edge from a satellite technique known as passive microwave,[27] Arctic-wide maps of sea ice thickness can be produced (e.g. Figure 4.3). This has only been possible since the launch of the CryoSat-2 radar altimeter satellite, which provides unparalleled coverage of the Polar Regions.

Airborne radar

Since its launch in 2010, Operation IceBridge has been valuable to the scientific community because it has secured long-term measurements of important study areas and has provided key data to serve research goals. By producing yearly measurements over land and sea ice, it has aided research in snow and firn studies,[28] ice sheet topography,[29] sea ice thickness,[30] glaciology[31] and more. The four Operation IceBridge (OIB) radars – accumulation radar, Ku-band radar altimeter, radar depth sounder and snow radar – were developed by the Center for Remote Sensing of Ice Sheets (CReSIS) at the University of Kansas. The radar depth sounder, for example, has been used to produce the latest

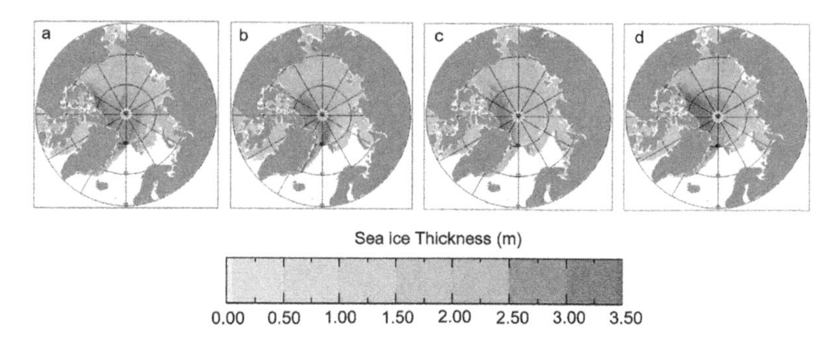

Fig. 4.3 Arctic sea ice thickness measured from the CryoSat-2 satellite, for spring (March/April average) (a) 2011, (b) 2012, (c) 2013 and (d) 2014 (Source: author).

ice thickness maps for the ice sheets of Greenland and Antarctica.[32] IceBridge data are publicly available through the National Snow and Ice Data Center (NSIDC): https://nsidc.org/data/icebridge/

Ground-based radar at Store Glacier, West Greenland

The phase-sensitive FMCW radar was recently configured as an experimental imaging system and deployed at Store Glacier in West Greenland as part of the SAFIRE project led by the Scott Polar Research Institute, University of Cambridge.[33] A photograph of the system deployed on the surface of the glacier is shown in Figure 4.4. The blue boxes contain an array of 8 transmit and 8 receive antennas which are arranged orthogonally to form a square 64-element virtual array looking down into the ice as illustrated in Figure 4.5. To achieve this, the radar signal is sequentially switched between 64 different combinations of transmit and receive antenna pairs.

Instead of emitting a single pulse at one designated frequency, as is the fundamental principle in traditional radars, this FMCW radar works by transmitting a chirp signal whose frequency is linearly swept from 200 MHz to 400 MHz. A wide signal bandwidth increases the vertical range resolution of the radar, in this case 42.5 cm with which individual layers within the ice can be distinguished. Conventional radars use high transmitting power to overcome the attenuation of the radar signal through the ice. Recent developments have allowed the peak output transmitting power level of the phase-sensitive FMCW radar to be as low as 100 mW, while consuming only 6 W of battery power during operation – less than a household light bulb.

Fig. 4.4 Photograph of the experimental imaging radar deployed on Store Glacier in 2014 (Photo: T. J. Young).

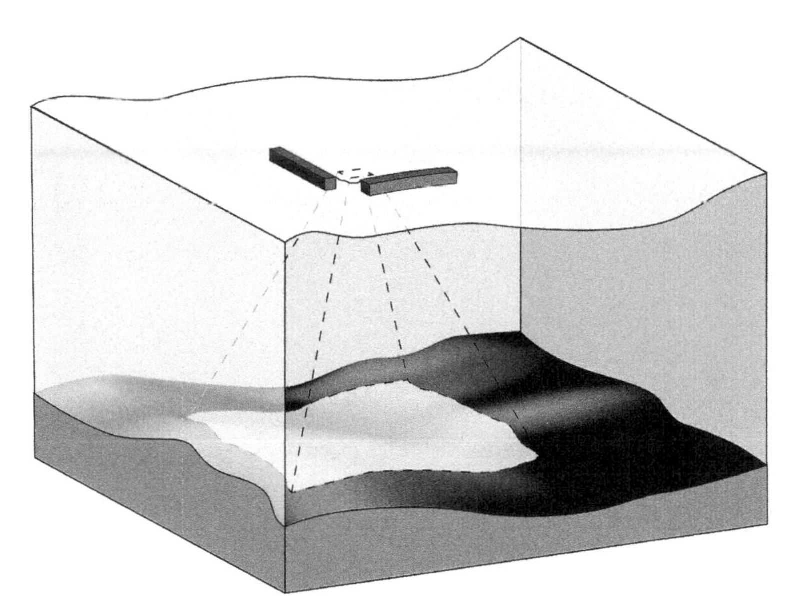

Fig. 4.5 Illustration of the experimental imaging radar operating on the ice sheet surface (Source: author).

Chirp signals are coupled into the ice using suitable antennas. Over the octave bandwidth of operation, certain types of antennas have proven to be useful and reliable. These include skeleton-slot panel antennas[34] and low-cost custom designed cavity-backed bowtie antennas[35] for long-term unattended operation. The very low power requirement of phase-sensitive FMCW radars provides a distinct advantage from a user's perspective. A static ground-based system that moves with the ice flow also enables many radar waveforms to be collected and coherently averaged to reduce the background noise level and therefore increase the detection capability of the radar. This advantage has been used to detect the base of grounded ice sheets over 3 km thick. [36]

A cross-sectional image through the ice formed by post-processing[37] the radar dataset is shown in Figure 4.6. The colour scale represents the intensity of the reflected radar signal. The processed radar image is analogous to an ultrasound of the ice. Here, three distinct layers of ice can be observed: an initial layer from the surface to around 100 m, followed by an ice layer 100–500 m, and finally another ice layer 500–600 m to the bed at approximately 618 m depth. These layers compare well with seismic data gathered in the same area and highlight the transition of the ice layers. At this particular site, a column of strong radar reflection was also observed from the surface down to around 100 m depth. It is believed this could be a water-/air-filled crevasse feature.

One problem with this experimental imaging radar, however, is the apparent curvature of the ice layers particularly from the bed – which should be horizontal in practice. The curvature is believed to be

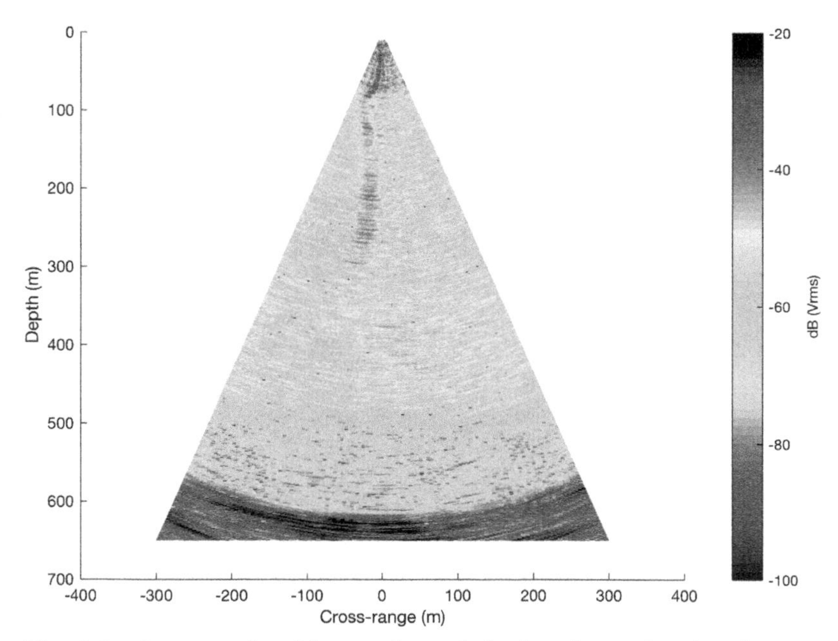

Fig. 4.6 Cross-sectional image through the ice after radar signal processing (Source: author).

a processing artefact resulting from the strong reflections that come from directly below the centre of the radar array, which swamp the true (weaker) returns that arrive from the wide angles. Improved methods of radar signal processing will be required to minimise this problem and this is the subject of further research.

These preliminary results from one field site in Greenland serve to illustrate the capabilities of modern radar techniques that are now being investigated further to help understand the processes occurring within ice sheets. The very low power requirement of the system means that it can operate all year, at hourly intervals, with a single 100 Ah battery. It can be complemented with invasive techniques such as borehole drilling to deploy sensors *in situ* within the ice.

Arctic science communication and outreach

The examples given in this chapter illustrate how radar can play an important role in documenting and encapsulating present-day climate change. For this reason, it is crucial to accurately translate the results generated by radar into accessible and engaging information to facilitate effective knowledge transfer to other academics, journalists, politicians and the general public. While it may be assumed that climate change policy relies on the latest results generated by science, there is a general reluctance from scientists in other disciplines, driven by various countervailing forces, to engage in political and public dialogue.[38] However, such a dialogue is a vital step towards efforts to understand and combat change in the Arctic. This section will address how to effectively communicate Arctic science to a non-scientific audience.

The 'dos' and 'don'ts' of science communication

Based on the experience of communicating our work to the public, press and politicians, we have compiled a list of the key 'dos' and 'don'ts' when answering questions or preparing a press release relating to scientific research.

Dos:

- Do communicate the scientific facts, but without any 'spin'. People tend to react negatively if it seems that you are telling them how they should feel about a certain result.
- Do speak or write in concise sentences that can be used as 'sound bites' for radio, or quotes in print.

- Do remember that context is key. For example, a statement such as 'CryoSat-2 measurements showed that the volume of summertime Arctic sea ice increased by 40% in 2013' relating to a paper[39] is highly ambiguous. Without any timeframe, it is not clear if this increase is relative to the last few months, years, decades or perhaps even longer. A much better way to communicate the same result would be to say 'CryoSat-2 measurements showed that in summer 2013, the volume of Arctic sea ice increased by 40% compared to the previous year'. This statement is far more difficult to misinterpret, and shows that although the increase in volume was large, it did not offset the loss observed over the past few decades.

Don'ts:

- Don't use any 'jargon'. Although everyone is an expert in what they choose to do, we are not all experts in the same thing. For example, at a conference one might say 'I use a 13.5 GHz satellite radar altimeter to measure the freeboard of Arctic sea ice' while for a non-expert audience this could be changed to 'I use a satellite to measure the difference in height between the Arctic sea ice surface and the water in the cracks'.
- Don't try and answer a question if you're not sure of the answer to it. It is acceptable to admit that you don't know, or that someone else may be better placed to answer.
- Don't be afraid of silence. Take time to consider the best way to frame your answer.

Outreach

The UK Polar Network regularly advertises opportunities for Polar Researchers to present their work and share their experiences with local school children.[40] Another common practice nowadays is to use social media to publicise one's research. The latest scientific results are often announced on Twitter within hours or days after a publication is released into the public domain. At universities, seminars and lecture series are another popular way for work to be disseminated publicly. An excellent example of this method was delivered by the late Dr Katharine Giles in 2012. It can be viewed on the YouTube archive[41] and contrasted with her original scientific publication. [42]

Conclusions and outlook

Radar technology has already allowed us to investigate vast areas that would otherwise be unreachable for humans, whether it be within an ice sheet or on the most remote Arctic ice floe. The amount of information that can be extracted from radar data is still in a nascent stage, and further advances in radar signal processing and data interpretation will help to unlock more insights from within and beneath the ice.

The different radar observation methods described in this chapter complement one another, as well as other geophysical measurement techniques described in the literature. Data storage and processing requirements will undoubtedly increase, along with improvements in computational power. It is now common practice for large volumes of raw datasets to be curated in digital format for independent validation and future (open) access. There is significant potential for improvement in battery and fuel technologies, in terms of their efficiency and weight, and future radar instruments can stand to benefit from advances in these areas.

As scientists who hail from outside of the Arctic, it is the remoteness of the regions studied and the nature with which they form, change and adapt under a changing climate that embodies scientific 'Arcticness'. But emotionally the Arctic is so much more – and is not necessarily remote. The beauty of the Arctic cannot and should not be described by science alone. And so the mystery of its scientifically unexplored and most remote regions continues to entice scientists, not just to study it from a distance, but also to visit the Arctic and to dispel all logic to embrace its magic and its peoples.

5
Arcticness: In the making of the beholder

Patrizia Isabelle Duda

Over the past 20 years, the Arctic has re-emerged as a space that is of geopolitical interest to regional and global actors alike. This development is particularly attributed to the Arctic being at the forefront of environmental concerns and therefore featuring prominently in international scientific, political and popular discourse. But the iconised melting of the Arctic is also portrayed in the light of alleged new opportunities. Extractive industries look to the potential profits to be realised by tapping into Arctic mineral resources; shorter Arctic shipping routes, which promise to cut time and costs, are tempting international shipping conglomerates; and the tourist industry promotes the Arctic as Earth's last frontier, making it accessible to 'ordinary adventure' travellers, who wish to experience it 'before it's too late'.

Alongside these developments, a vague notion of 'Arcticness' is appearing throughout academic and political discourse. While Arctic national strategies make direct or indirect references to their (sense of) Arcticness, others wonder whether Arctic nations and states have successfully shown their Arcticness.[1] Yet not only Arctic, but also some near-, sub-, and non-Arctic countries position their interests in the region by arguing for their Arcticness on the basis of geographical proximity or security and economic interests[2] while asserting their 'perceptions and strategies of Arcticness'.[3] These perceptions and strategies then suggest the complexities of the term Arcticness: namely, its contextual meaning and usage in accordance with the different perceptions and strategies of a range of societal, political, economic, environmental and scientific actors.

Perhaps a good place to start in order to shed some initial light on this notion is by looking at the most obvious: the popular perceptions and fantasies of 'outsiders'; that is, populations living outside of the Arctic. Of course, this in itself constitutes a problem of definition. When is somebody considered to be living 'outside of the Arctic'? Does that refer to any point below the Arctic Circle, or are people in close sub-Arctic regions with similar conditions still 'too Arctic' to be considered non-Arctic? Furthermore, are southerners' perceptions of so-called Arctic nations different based on an assumed, more intimate understanding of their countries' northern dimensions?

Here, a simplistic yet pragmatic line is drawn by referring to 'outside perspectives' as being those of people who do not belong to Arctic countries defined geographically. While it may seem trivial to explore these outside perspectives in this context, they are far from irrelevant as they have the potential to influence Arctic politics and, hence, 'insider' narratives of Arcticness. One glance at current environmental debates suffices to reveal the power of outsiders' perceptions and attention. Hence, exploring these outsiders' perspectives is well worth the effort.

It appears that even among societal elites and non-elites who have perhaps never experienced the Arctic first-hand, there seems to be an almost intuitive feeling of what Arcticness might be. The examination of popular culture, literature and visual arts (see for instance, Corey Arnold's Arctic photography: http://coreyfishes.com) shows that particular Arctic attributes dominate the popular imagination:[4] snow, ice and cold weather; untouched, pristine and desert-like landscapes; long, dark periods versus long periods of eternal sun; indigenous communities; fishermen (this gendered term is used here intentionally) and masculinity; the intimate relationship between people and their environment, as well as an abundance of natural resources – or in other cases, the scarcity of resources.[5] In sum, the Arctic, the north, and the apparent Arctic-north construct in the outsider's collective image resemble a near-binary dichotomy of simplistic generalisations and stereotypes of nature and culture(s).

While popular media on climate change concerns may contradict notions of the Arctic as a pristine region,[6] the efforts to raise awareness of its global relevance beyond Arctic latitudinal borders[7] rarely go beyond reminding viewers of the fragility of this space through the threat of near apocalyptic determinism. As such, they reinforce some of the listed fantasies about the Arctic and the perception of a global imperative to protect it. That is, Arcticness emerges as a collection of the same aforementioned environmental and physical attributes that are now simply of a more global concern, driving outsiders' calls to keep the Arctic as the untouched, pristine place they perceive it to be.

Seen this way, Arcticness represents the character of the Arctic as featured in outsiders' collective constructs – a more or less coherent space disrupted by climate change and made up of the aforementioned elements or a perceived absence thereof, rather than a more realistic, 'dynamic, transnational, connected and contested region where natures, identities, histories and politics all intersect'.[8] In fact, the distance between outsiders' perceptions and 'northern peoples' perceptions of themselves and their homeland is as vast as the Arctic landscape'.[9] The analysis of such 'at best simple and incomplete and at worst incorrect and prejudiced perceptions',[10] have led some scholars[11] to draw connections to Said's 'Orientalism'.[12]

The resulting references to 'Eskimo Orientalism'[13] or 'Arctic Orientalism'[14] are not without merit as collective images of the Arctic have been fundamentally driven by either past or modern scientific imagery.[15] More so, these served as a mirror of one's (western/southern) self, 'a strategy of imagining the self as an explorer-hero, a scientific worker, or a white, imperial male'[16] – ultimately emphasising the (political) power to objectify the 'other' versus the civilised 'self'. Thus, it is perhaps unsurprising that colonially charged male Arctic exploration and scientific inquiry, and the resulting descriptions/travelogues, have contributed to gendered portrayals of Arctic nature and indigenous communities and cultures and, thus, perceptions of Arcticness. This gendered image is further perpetuated through influential popular magazines such as *National Geographic*,[17] high literature and visual arts, taking inspiration from such 'scientific' imagery. Arctic nature is then 'romanticised through literary stereotypes based on masculinist values'.[18] Consequently, manhood-related themes such as wilderness and breaking away from home, courage and heroism, domination over infertile land, of man over nature, and with it the conquest over 'uncivilised' communities and exotic indigenous females became defining features of outsiders' perceptions of the Arctic.[19] Remarkably, until today, 'alternative ways of perceiving northernness are extremely rare'.[20]

Yet, while our collective Arctic image may have remained constant, others argue that it has slowly begun to change since the mid-2000s as a result of embracing new, often climate change-related concepts that generated more nuanced perceptions.[21] Of course, 'new' does not necessarily mean more gender-neutral. The obvious portrayal of heroic masculinity alluded to here seems less salient and less socially appropriate today, but our political and to an extent, scientific institutions still have concrete foundations based on such images of masculinity. They may dictate the type of personnel expected to reach and thrive inside the Arctic Circle as well as their likely behaviour. We can refer to the better-studied topic of military systems and quote Dixon: 'The argument is simply that a proportion of those [individuals] who opt for a career in the armed services ... will be attracted to organisations which set them upon the seal of masculinity ... being admitted to a society of men bent upon the most primitive manifestations of maleness'.[22] Can we then expect that Arcticness is still affected by the potential of an assumed male prevalence in these realms?

Interestingly, some Arctic players are further reinforcing these outsiders' images in the effort to capitalise on increased global interest in the Arctic – for instance, through national branding initiatives. Iceland serves as a prominent example. Its tourist and state industries have successfully reproduced these images into a unique national brand that fits the globalised neoliberal world's commercialisation of distinct cultures, ethnicities, and exotic and isolated places.[23] Thus, outside notions of Arcticness are reinforced by advertisements of the Arctic region as, for example, a moon-like, empty last frontier, by the selling of €9 boxes of 'fresh Icelandic Mountain Air' in Icelandic tourist shops, Rovanemi's (Finland) branding as the official home town of Santa Claus, and by cruise brochures' shiny images promising vast landscapes of ice, polar bears and the aurora while recalling the authenticity of past explorers' experiences. And what historically was often seen as a problematic, 'uncivilised' hinterland is now embraced by all sides – after all, indigenous people performing 'unusual' rituals and traditions are well-suited to staging the sought-after authenticity.[24]

Consequently, perceived Arcticness is also manifested through the branding of the Arctic as a pristine, exotic, gendered, wild and uncivilised place[25] to attract adventurous, 'independent' tourists, enthusiastic to explore the allegedly 'undiscovered' and rough character of the Arctic, turning 'their holiday tours into a mode of exploration and their narrative personas from tourists to adventurers'.[26] But, as Loftsdóttir's analysis of Icelandic national branding efforts demonstrates, to succeed, such campaigns require 'already existing stereotypes and conceptions',[27] filtered through the colonial past 'into the present, shaping contemporary global imaginings of difference'[28] that 'emphasise the association with the exotic, from which Icelanders had tried for so long to distance themselves'.[29] Hence, perceived Arcticness serves as a branded commodity in a global marketplace. Arcticness represents, among other aspects, characteristics considered lost by so-called 'advanced' regions[30] which are valued particularly by younger generations who are influenced by neoliberal ideals.[31]

However, it is not merely 'economic actors' who are 'selling out' by defining and using what might be considered 'ignorant' outsider notions of Arcticness in order to make economic gains. Similar 'marketing' processes take place in the political realm, although with different goals and purposes: those of international strategic positioning. While an ostensibly descriptive and overarching term such as Arcticness may indicate certain regional belonging based on a more or less uniform set of national characteristics, this is not the case.

In line with increased popular, economic and political interests in the region, all eight Arctic sovereign states have published comprehensive national Arctic strategies, manifesting their belonging, legitimacy and interests in the Arctic.[32] Most 'found their "Arcticness" only after the publication of the ACIA report in 2004'.[33] Examining these strategies shows that all involved parties justify their Arcticness and, hence, their self-appointed legitimacy for tapping into the region based on partially differentiated sets of reasons that fall primarily into five categories: security, sovereignty, environmental protection, social and economic development as well as the governance and administration of the region. How these categories are then integrated into each nation's Arctic strategy, and their claims of Arcticness, depends on their unique geographical, economical, technological and historical starting points.[34]

Consequently, with uncertainty still prevailing over the region's tangled interests, institutions and future political constellations, carefully disguised power games may be the name of the game. That is, Arctic narratives emphasise Arctic international politics as essentially being characterised by peaceful cooperation among all the stakeholders.[35] The motivation for this stance is to subtly establish de facto power in anticipation of future economic and (to follow the (neo)realist line of thought) territorial gains and any potential disputes over them. The Arctic can thus be portrayed as a set of actors, interacting in ways that demonstrate different, more integrated ways of undertaking international dialogue and politics.[36] One illustration of integration is the establishing of (some 'unlikely') state partnerships and non-state actors being brought to the institutional 'Arctic Council table' as equals. The Arctic Council table is supposed to take a lead position in enabling its members to cooperate and shape policy-making.[37] Does this point to Arcticness as a new, more cooperative, inclusive and peaceful approach to international politics, be it based on tactics or even on previous international political lessons?

To some degree, this seems to be the case, considering the 'unparalleled level of indigenous political engagement' in high-level politics.[38]

Moreover, there are collective historical memories of a challenged and divided Arctic region,[39] most recently by the Cold War – an issue which to some extent still exists due to ongoing NATO divisions. Such memories of division provide a powerful impetus for public narratives to emphasise the necessity of peaceful cooperation for the secure governance of this environmentally harsh region. This is especially true when much international relations history and literature point to the region's potential for conflict.[40] However, this narration of space, be it by outsiders, economic actors or states, may have exerted significant influence on states' foreign policies as a visibly more cooperative approach has been developed since the 1990s aided by the institutionalisation of cooperation on non-military matters through the creation of the Arctic Council in 1996.[41]

While Arctic states portray peaceful cooperation as part of their Arcticness, the atmosphere seems to change when non-regional Asian and European middle powers, China, or the European Union (EU) as a soft or smart power, express a wish to join the Arctic club.[42] Ironically, these non-regional nations construct and 'sell' their own Arcticness and, thus, their legitimacy to the 'Arctic pie' as a global common.[43] Those who have successfully wooed the Arctic Council are now acting as observers in cross-cutting middle power diplomacy. They are 'bridging public diplomacy' and creating niche alliances on many specific subjects, environmental issues and so-called 'green growth'. But they are also entering some traditional domains of big power politics in order to exert their influence on Arctic agenda setting.[44] A focus is thereby put on soft areas of technical win–win situations (i.e. ship-building technologies, investments in Panamax ports, cooperation on expensive scientific projects and environmental technologies), demonstrating 'southern solidarity', reiterating shared experiences[45] or any issue which is regarded as essential to the Arctic. In this forum, the EU is something of a special player. Its desire to be involved is not only motivated by Brussels' energy and environmental concerns but also by the geographical position of its northern member states within the Arctic.[46]

As alluded to above, these outsider efforts are often met by the Arctic club with some nervousness and the use of stalling tactics. This has been demonstrated most recently by the 2015 deferment of allowing the EU observer status on the Arctic Council. The continued mistrust and reluctance of Arctic countries to fully include non-Arctic states points to some elemental characteristics of what Arcticness means to the Arctic nations. Namely, it is not just a static identity-politics but it also involves region-building[47] via the social exercise of 'zoning' the Arctic as a distinct space with respect to other states and regions.[48] At the same time, the Arctic nations are asserting their own belonging to this socially constructed 'imagined community'[49] in relation to the external 'other'. The external other in this case can mean both positioning oneself 'with the other' as is currently the case vis-à-vis those nations' Arctic 'hinterland' or other regional states, and positioning oneself 'versus the other' as is happening vis-à-vis non-Arctic players. Interestingly though, efforts among Arctic states to establish their Arcticness based on some common features of regional belonging are almost immediately followed by an emphasis on the differences between them. This lends a unique colour to their individual Arctic claims for legitimacy to be part of their own club.

Establishing one's Arcticness, therefore, is built partially, but powerfully, by utilising the notion of a margin as a tool or considering the dichotomy between centre and margin, with which to side or which to block to naturalise these social constructions until they become accepted as reality. Hence, Arcticness means an exclusive club of states deciding who to include and cooperate with and who is not worthy or trustworthy enough to be granted this status. By the same logic, non-Arctic players construct an Arcticness in their national identity narratives that rationalises their envisioned involvement in the Arctic arena through non-traditional justifications that deviate from ethnic nationalism and traditional power politics. In short, sub-Arctic actors assert their status as 'Arctic stakeholders with real rather than imagined stakes'.[50]

In conclusion, and at the risk of stating the obvious, Arcticness is not a static value but essentially the rich and dynamic processes of outward- and inward-looking imagination, identity-building and establishing belonging motivated by a plethora of reasons. In the narrower scope of states' Arcticness, it is just as Rostoks[51] writes (drawing on Wendt)[52] 'The Arctic is what States make of it'. So is Arcticness what states make of it: a social construction of identities, interests and power politics in an arguably anarchic system? With political debates once again turning to competing narratives of human nature as projections of the Arctic's future fate – split between (neo)liberal assumptions of cooperation and (neo)realist warnings of looming conflict and hostile competition – a more balanced view is sometimes lacking. Yet it takes only an innately 'empty' concept such as 'Arcticness' to demonstrate how a social construct and related discourse becomes an instrument of power politics in a state's toolbox.[53]

In the wider context, it is the shared ideas and interactions of different actors that over time have given meaning to the notions of

Arcticness, in fact creating them to begin with. Arcticness then emerges from the perceptions and resulting collective images of outsiders, built upon backward-looking sets of historical legacies and forward-looking alleged threats and opportunities driven by economic, political, technological and environmental factors. Environmental factors in particular have had an enormous effect on Arcticness being characterised as an integrated, distinct region, seen as 'either an ecologically protected space or as a space of natural resource exploitation'.[54] This puts the Arctic on the map again and contributes to northern countries' empowerment due to their potential to be a local resource rather than just a global environmental problem. Finally, Arcticness is powerfully affected and effected by efforts to utilise or manipulate the common perception based on globalised (neo)liberal values and fantasies, especially among younger generations. Hence, Arcticness *is becoming* what states and actors make of it. At heart, it is an issue of identity, power and interest-formation.

PART 2
Arcticness Living

6
Arcticness insights

Anne Merrild Hansen

What is special about people living in the Arctic? Do we have more in common with other Northerners than we do with people living south of the Arctic Circle? What makes us what we are? These are the types of questions I started wondering about when I was confronted with the term 'Arcticness' in relation to the creation of this book. Being a Northerner myself, and based on my experiences as an Arctic researcher conducting fieldwork across the region, my view is that there is a special bond among Northerners and also particular values and interests that we commonly share.

I believe that the Arctic environments shape the lives of its peoples, the traditions, views and livelihoods. Dark winters and light summers, remote settlements and sparse resources are features that bring the communities together. But there are also great differences. While people in Barrow, Alaska, are living on the open frozen tundra, Greenlanders are living in coastal areas surrounded by mountains and Saami nomads are crossing large distances inland every year. Even though there are common challenges related to living in the Arctic, the environments in the different countries vary and influence peoples' ways of living.

Other things besides the environment influence and frame the life of Northerners. Human decisions on local, national and international scales and actions throughout history influence the way we live and perceive ourselves. The perception of what an Arctic identity is and entails is therefore as unique to communities as it is to individuals. In this chapter, I focus not on the differences but rather on the common characteristics of Arctic peoples, the Northerners.

Based on my biased expectation of an Arctic identity being a reality and to reflect on the potential characteristics and to gather inspiration for this chapter I reached out to my cross-Arctic network through my personal profile on social media (LinkedIn and Facebook) and asked my connections what they find is special about being Northerners. I particularly encouraged my contacts from the Arctic region to finalise the sentence: 'You know you are from the Arctic when ...?'

More than forty responses were posted within a few days, from people living in Russia, Norway, Iceland, Greenland/Denmark and Alaska. A few asked me to contact them privately and this led to interesting and good conversations on the topic of Arcticness. As expected, most responses came from Greenland where my Arctic network is widest.

Not claiming the results from this small social media exercise to be representative in any manner for the opinions of Northerners in general, I still find that they point to general characteristics, which I will share here as they offer insights into how we as Northerners see our own reflections in other Northerners. The responses were, probably, partly due to the publicity on social media and partly due to the populist way I formulated the question, written with a focus on positive features and not on negative perspectives of living in the Arctic. There was a romantic tone in many of the replies and not a single sarcastic reaction. These inputs from my Arctic connections supplemented and nuanced my own reflections. I found them both funny and thoughtful and I have to a large extent used them in the following. I do not claim to point at cultural markers and I would never dare to try to define what identifies a Northerner.

The chapter is meant merely to present a snapshot in time of personal perceptions by fellow Northerners and myself on our common characteristics in 2016. The topics in the replies were inherently interconnected, but for the sake of simplicity I have grouped them under three headings, which I present and elaborate on.

The sounds of quiet

The environment in the Arctic is often described in international literature as fragile, vulnerable and sensitive, but as a Northerner you tend, rather, to perceive the surrounding environment as great, strong and potentially dangerous, fostering respect and continuous adaptation. The weather conditions are extreme and harsh, but nature is also the provider of the resources needed to survive and the greatness and beauty of Arctic nature is stunning to its residents. I expect that this is a part of the explanation as to why we as Northerners feel need, love and fear of Arctic nature all at the same time.

Emphasised by my Arctic contacts, in relation to Northerners' connection with nature, is the love of the sounds of quiet. 'We love the greatness of our nature, which allows the remote feeling that gives you room to breathe and space to unfold' as one of my Greenlandic friends stated. The silence is particularly remarkable in the Arctic as it contrasts the noises whenever a storm or blizzard sets in and the moments of silence are the times where you stop and think, and make you present in the presence. One of the respondents noted 'You know you are from the Arctic when you live in nature and the nature lives in you'. And this underlines the special bond Northerners feel to nature.

Related to the silence and the bond to nature, a particular Arctic phenomenon is the aurora borealis, also called the polar lights, but by Northerners mostly known as the *northern* lights. The magnificence of the bulging light waves across the skies at night is subject to various myths and legends across the Arctic region. As a child growing up in South Greenland, I was scared to whistle during northern lights, as it was said that the light was created when the dead played ball across the sky, and if you whistled, they would come and take your head to use it as a ball. In general, ghost stories and creepy myths are also something we share in the Arctic. I think the huge wilderness and uncontrolled nature, and the long dark winter nights, create an atmosphere that invites these stories to be told. Every place I have visited in the Arctic, people have their own stories and love to give each other the creeps. As noted by a woman from Alaska, you know you are from the Arctic 'when you grew up with stories about the little men of the tundra and their poison tipped spears'.

In relation to the natural environment and climate in the Arctic, an inevitable part is the cold and the fresh air and low humidity. Related to the sound of silence and the cold and dry climate, therefore, is the sound of squeaking snow. Snow, when it is really cold and dry, makes a particular noise. Some years ago, I conducted an interview with a lady who was, at the time, the oldest woman in Qeqertarsuaq, Greenland. I was asking her about issues related to climate change and when I asked if the weather in recent years was good or bad, she answered that she could not define on behalf of others what was good or bad. What may be good weather for fishers may be bad weather for hunters, but one thing that she could tell me was that she missed the sound of squeaking snow under her kamiks (seal skin boots) in the winter as the temperature rose and the winters became milder. The interview was one of those that left an impression, and I clearly remember her face and voice.

Isolation and togetherness

Arctic communities in general are small in size and the number of inhabitants is low. The communities are typically geographically dispersed and remote. The infrastructure is sparse and travel happens by plane, boat, helicopter or snowmobile or even by sledge, on horseback or on ice roads in the winter. This means that many communities are as isolated as if they were each located on their own islands, creating a situation where social relations among residents within a community can be very strong and intimate.

Relations between different communities are also important. In Greenland, one of the first things you ask when you meet other Greenlanders is who they are related to. While I was conducting interviews in North Greenland in 2013, all the conversations I had were initiated by the interviewees asking about which town I come from and about my background, parents and relatives. When they had an idea of where I belonged and common acquaintances were identified, the interviews could begin. Family bonds tie people together and a comfortable atmosphere can be reached when relations are shared, and they always are. We are so few, that family and friends are always to some degree shared within the same countries, even when we each live in our end of the country.

Another aspect related to living in the small communities in the Arctic is that we live close to each other. As one stated: 'You know that you are from the Arctic when grocery shopping involves talking to friends, neighbours, and colleagues'. It is difficult to hide abuse, an affair or crime. And even though taboos exist here and there are things that are 'not being talked about', secrets are not easily kept. Issues connected to the social relations and networks in the communities were highlighted by Northerners addressing the question of what being Arctic entails, with an emphasis on how we, to a large extent, accept peculiarities of other community members, not least in recognition of every person being a resource in a local community, and in recognition of most people or families having their own secrets to live with and for others to accept.

Social relations and togetherness are widely practised and Northerners are generally open and welcoming. Get-togethers involve a lot of eating, often including sweets and cakes along with local delicacies such as fermented shark, sheep, raw whale skin or special cuts of moose or bear meat (Figure 6.1). Storytelling and sharing at gatherings contribute to upholding what we perceive as a good quality of life.

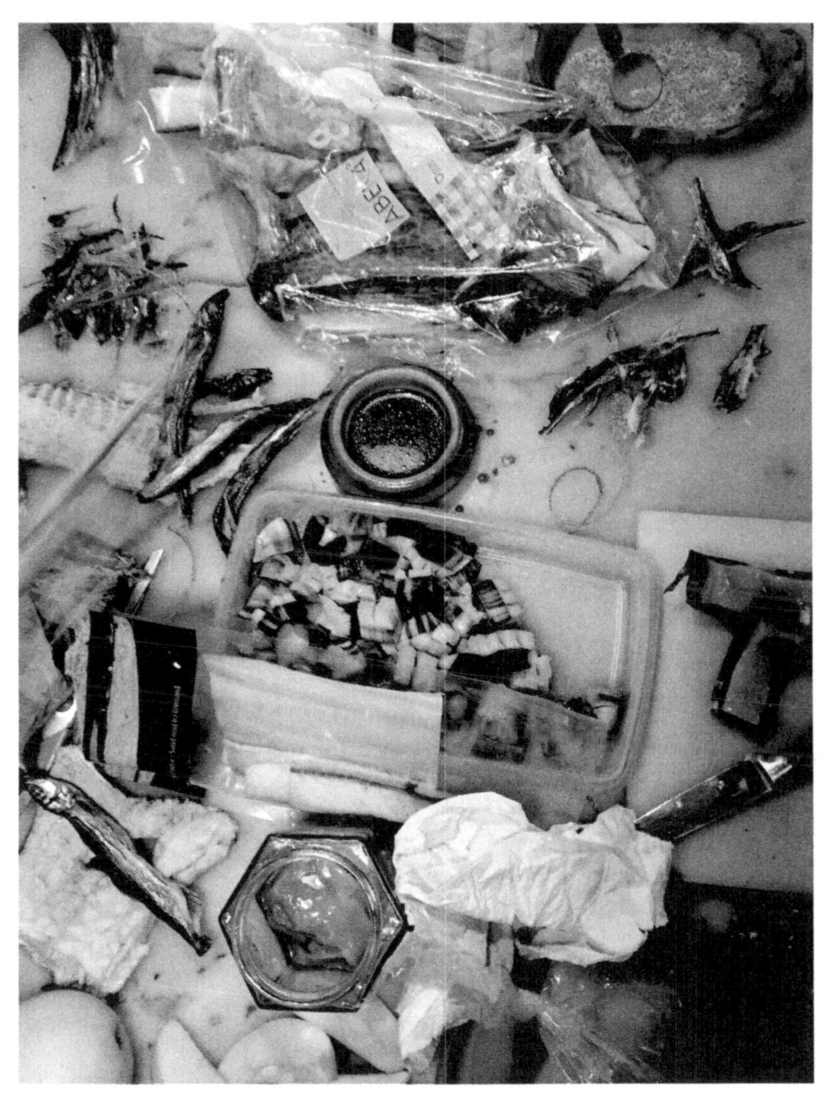

Fig. 6.1 *Mattak, panertuut, iginneq* and other delicacies from Greenland (Source: author).

Meat we eat

The selection of goods in Arctic communities is often limited. Natural fresh food resources such as muskox, caribou, seal, fish, berries, seaweed, sea birds, mussels, mushrooms, whale and moose are free and available, so subsistence hunting and fishing are activities of great importance to Northerners. As the vegetation is insignificant due to the cold, the storms, permafrost and rocky ground, greens are hard to grow. Meat is therefore an indispensable part of the Arctic diet.

Limited infrastructure and long transport distances mean that supplies are often expensive and sparse. Self-sufficiency is an important contribution to Arctic diets. Even in the larger Arctic towns where the stores can provide most goods, hunting, fishing and gathering are still considered essential for well-being and are practised both for recreational purposes and to supply food.

As hunting and fishing are such an integrated part of living, access to relevant tools is essential. Several responses I received from my Arctic network about Arcticness were about how easy it is to access weapons and how common it is to carry and use a rifle. One emphasised that 'You know you are in the Arctic when you can walk into a bank with a rifle on your shoulder and not be arrested' and another said that you know you are from the Arctic 'When you can walk into a store and buy a rifle on special offer; but milk won't be in till at least Wednesday (true story)'.

People in the Arctic have a strong bond with nature and time is spent in the wild; hiking, skiing, climbing and sailing. And the cold does not bother Northerners. As one stated, 'When the temperature at the lake rises to +5°C, children start to swim'. We also know how to dress in the cold. Clothing made of leather, fur and wool is commonly used and as one stated: 'You know you are from the Arctic when your date wears long wool underwear even in the summer'. Dressing warmly is something we naturally prioritise, and we learn to dress in layers. During time spent on the land, meat is gathered, which is why you often see animals and fish being cut outdoors; chunks of meat hanging outside buildings to tenderise, or lying spread on rocks for drying, or hanging on racks. Skins are similarly being treated by hand and hung to dry. Butchering is a skill that is passed on to the younger generations and blood is an inherent part of this. Northerners therefore have a relaxed attitude to killing and butchering and one of the responses underlined this with the statement: You know that you are from the Arctic 'when you think people are strange who get sick by the sight or smell of blood'.

The interconnectedness of characteristics and perspectives on quality of life in the Arctic

As described in the previous sections, the Arcticness characteristics highlighted all relate to each other (Figure 6.2). The love of silence is

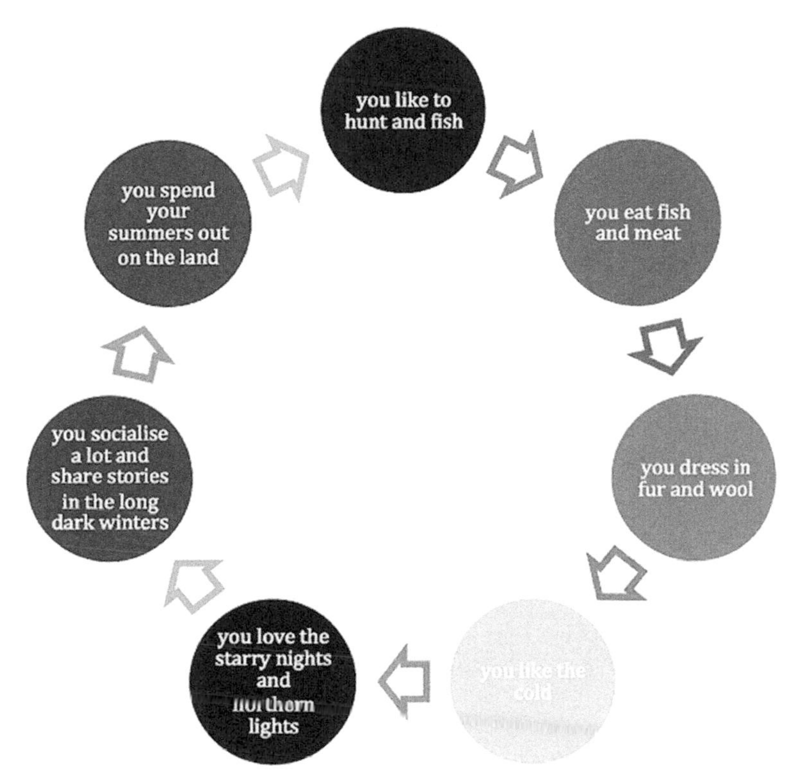

Fig. 6.2 Interconnected characteristics: Northerners according to Northerners (Source: author).

related to the love of the land that provides the meat we eat and the way we live in remote communities where there is room to be different and still be a part of the community. The Arctic environment offers the frames and the resources in which identities, traditions, norms and values of Northerners are developed; frames which are to some extent similar and resources which are similarly rich in some ways and similarly sparse in others.

My personal experience is that the common frames and resources mean that it is easy to be in the company of other Northerners. We laugh and cry over the same things and share similar experiences and in this way recognise our own reflections in each other. That is why we feel comfortable and understood by our Arctic neighbours to a larger extent more than we do with others.

In everyday life, we seldom stop and consider what it is that makes our life what it is. We are not consciously reflecting characteristics or values, but the values present in our life make sense and fill it with what for us is the prerequisite for quality of life. Returning to the question of what Arcticness entails, I think that the overall main characteristic of a Northerner is a person who loves and thrives in the Arctic. Northerners live in the Arctic not because they have to, but because they want to.

7
Reindeer herding in a changing world – a comparative analysis

Marius Warg Næss

Introduction

Imagine for a moment that you wake up one morning: Getting out of bed, you look out of your window and discover that a lot of snow has accrued during the night. You start to panic: how will your livestock do in this weather? The snow is too deep for them to dig through to find fodder and they are therefore at risk of starving to death. Luckily, you can move your herd to another pasture that you have saved just for such an emergency: To get there, you need to move through pastures that have been used by your neighbours and collaborators for many years. On the way you discover a newly erected fence that stops you dead in your tracks. At the same time one of your former collaborators, quite angry, tells you to turn around. He says that you cannot move on because this is now his 'private' pasture area not open for anyone else.

While a somewhat caricatured story, strangely enough it is a description that fits the situation currently facing herders on the Qinghai-Tibetan Plateau[1] and might as well be the future for reindeer herders in the Arctic parts of Norway. On the Qinghai-Tibetan Plateau, re-allocation of grazing areas and fencing has been going on since the early 1980s[2] and has already resulted in war-like conditions. A dispute relating to grazing rights resulted in the deaths of at least 29 Tibetans between 1997 and 1999: starting small, the dispute soon escalated into periodic armed fighting, involving some 2,000 fighters using automatic and semi-automatic weapons.[3] In the Arctic, the Norwegian government is currently in the process of privatising previously semi-common winter pastures as this is assumed to be an important prerequisite for developing a sustainable reindeer husbandry.

As privatisation is currently happening in Norway, we do not really know how, if at all, it will affect reindeer herders. Nevertheless, a substantial amount of comparative evidence exists that can be used to critically investigate the current policy and its possible effect on reindeer herding, that is, developing scenarios for reindeer herding. Pertinently, the Qinghai-Tibetan Plateau has a cold climate and is covered by cold grasslands that are similar to the cold grasslands of dry-tundra regions of the Arctic,[4] making it a useful comparison. Scenarios are a way to envision possible futures and while they are sometimes understood as being a prognosis for the future, here scenarios are better conceptualised as storylines about how the future might unfold.[5]

Comparative aspects of land tenure privatisation

In general terms, nomadic pastoralists have traditionally owned animals privately: rangelands have been owned – or at least regulated – informally by groups of herders. The underlying rationale for the privatisation of pastures is usually twofold: on the one hand it is driven by a desire to develop pastoral societies. In this light privatisation makes perfect sense because it renders pastoralists less mobile and thus enhances governmental objectives of providing basic social services such as education and health. Mobility has led governments to look at pastoralists as 'backward', lacking the technological level and skill to successfully exploit their existing adaptation. Thus, in many areas of the world large governmental sedentarisation programmes have been established to raise the technological level, and to enhance the profit of pastoral production.[6] But it also provides a form of governmental control lacking when pastoralists were constantly on the move – not only within sovereign national states, but also across state borders.

On the other hand, there has been an interconnected concern of sustainability: it is assumed that pastoralists are trapped in social dilemmas where individuals act independently and seek to maximise short-term gain to the detriment of collective benefits.[7] Hardin – with the introduction of the 'Tragedy of the Commons' (ToC) – provided a framework predicting that pastoralists would increase stocking rates to such a degree that overgrazing was inevitable; in other words pastoralists are 'overstockers'.[8] This implies that pastoralists are unable to establish rules and norms that minimise, for example, overgrazing:[9] it is a widely held belief that common ownership of land coupled with private ownership of livestock and the lack of a strong state provides incentives to degrade the environment.[10] Consequently, nomadic pastoralists have been viewed as non-rational, and professionals and governments have seen problems, such as pasture degradation, as *inherent* in the nomadic pastoral adaptation.[11]

Privatisation is thus occurring within an official debate pertaining to overgrazing and rangeland degradation. The debate in China is illuminating. There it is argued that increasing land degradation is caused by (1) increased livestock numbers (from approximately 29 million in 1949 to 90 million in the early 1990s) and (2) a decline in the area of available rangeland (around 6.5 million hectares were lost from 1949 to 1992[12]). Notwithstanding an apparent increase in livestock numbers, the evidence for degradation is somewhat tenuous: according to Harris,[13] in 1999 the State Environmental Protection Agency estimated that one-third of China's grasslands were degraded, but in a very short time the figure that is often cited increased to 90 per cent without any obvious scientific reason (generally, estimates of degradation in China have been based on varying subjective measures and have been poorly documented – no systematic investigation has been undertaken[14]). Similarly, in Norway the official policy is based on the assumption that fixed grazing boundaries are a prerequisite for establishing an ecologically sustainable upper limit on the number of reindeer and will serve as a facilitator for rational resource use.[15] In short, despite apparent differences in overall political systems, the decision to privatise pastures seems to be driven by a common ideology presupposing a ToC and overstocking in both Norway and China.

Land tenure

Land tenure can be defined as the relationship between people and the land, and the rules that regulate how the land can be used, possessed and redistributed;[16] or as the mode by which land is held or owned; or by the set of relationships among people concerning use of the land and its product. Land tenure refers to the societal institutions (organisations, rules, rights and restrictions) that control the allocation and use of land and its associated resources.[17] Generally, land tenure is often conceptualised as: (1) *commons* (common property) – land is treated as commons with no enforceable control over access to resources; (2) *reciprocal access* (communal property) – there is reciprocal access between members of land owning groups; transfer of group membership (the foundation of property right) is easily negotiated; (3) *territoriality* (local group ownership) – strong control on local group membership and a reduction in reciprocal access; and (4) *private ownership* – ownership devolved to well-defined subsets of local groups (e.g. kin groups or individuals).[18]

A chronology of land tenure changes in Tibet and Norway

For both Tibetan herders in China (*drokba*) and Saami reindeer herders in Norway, the basic unit of social organisation is the household, a nucleus or stem family. Traditionally, households often combined together and formed small cooperative groups that shared nearby pastures, called *ru skor* in Tibet[19] and *siida* in Norway.[20] In some parts of Tibet, *ru skors* were aggregated into higher order groups called *tsowa*.[21] The *tsowa* has been predominantly described for the east and was organised around a lineage of a particular founding patrilineal clan that controlled bounded tracts of land.[22] While the land rights of *tsowa* were fixed – unless and until other tribes took them by force – the rights of individual *ru skor* were fluid.[23]

In contrast, nomads in the central and western parts were all under direct state control.[24] In principle, all of the land in Tibet was owned by the central government in Lhasa, which distributed the land among the aristocratic families, great incarnate lamas and monasteries for their upkeep and support. The nomads had to pay taxes and provide labour services to the institutions; in return the lord had to maintain law and order.[25] Pastures were re-allocated at three-year intervals based on the herd size of individual households. Additional pastures were allocated to households whose herds had increased, and pastures were taken away from those whose herds had decreased.[26]

In Norway, the *siida* seems to have been the highest social aggregate, but following the Reindeer Law for Finnmark, from 1854 reindeer herding was formally (and physically) separated into different summer districts.[27] Winter pastures on the interior constituted an overlapping quilt due to an absence of physical obstacles and because they were less formally governed.[28] While pastures were technically Crown land, the *siida* formed the basis for user rights both within districts during the summer and on the winter pastures. In other words, the customary tenure system was based on *siida* user rights (albeit informal). While winter pastures were *informally* regulated according to *siida* membership – that is, Saami reindeer herders had a clear understanding of the fact that different winter pasture areas belonged to different *siidas* – when in need everybody had a right to access alternative pastures.[29]

In Tibet, the traditional system was effectively dismantled during a period of collectivisation. The Cultural Revolution – a campaign to destroy the 'four olds', that is, the old ideas, old culture, old customs and old habits – arrived in Tibet in the 1970s and almost destroyed the nomads' way of life.[30] While the pastoral technology stayed the same, ownership of livestock and decisions regarding production were transferred from the household to communes, the collective production units.[31] Under the traditional system, only the distribution of pastures was controlled by the state; after the Cultural Revolution all aspects of economic and social life were fixed by state policies. Pastoralists were the subjects of commune leaders, and received work points, or 'stars', for their labour. The work points became the basis on which they got food, goods and cash.[32]

The Saami herders in Norway never experienced anything as disruptive as the Cultural Revolution. Nevertheless, while both the *siida* and household retained their positions (the household in some sense became strengthened at the expense of the *siida*[33]), the traditional tenure system was dismantled with the 1978 Act. This Act introduced a system whereby the Saami own their herds while the rangelands – owned by the Crown – are administered by the Ministry of Agriculture through the Reindeer Herding Administration which plans and regulates the distribution of herds and the grazing time schedule.[34] The most disruptive aspect of the Act redesignated the autumn/spring and winter pastures as 'commons'. It has been argued that as the 1978 Act did not incorporate any system for managing the pastures, it effectively 'led to the exclusion of the customary tenure system and, in the absence of a functional alternative regime, created *de facto* a situation of open access to resources' (p. 215).[35]

In the 1980s the communes were dissolved in China and the Household Responsibility System (HRS) was introduced.[36] In short, the HRS re-established the household as the basic unit of production and management decisions were largely devolved to households. For pastoralists, the HRS was implemented in two stages: first the privatisation of livestock and second the privatisation of rangelands.[37] Since the dissolution of the commune system, Chinese government policies have emphasised that individual household tenure is a necessary condition for sustainable rangeland management[38] as well as increased production.[39] By the end of 2003 around 70 per cent of China's usable rangeland was leased through long-term contracts, where 68 per cent was contracted to individual households and the rest to groups of households or to villages,[40] although estimates vary.[41] Consequently, the *ru skor* seems to have been destroyed in the east,[42] while cooperative herding still occurs and provides a necessary component of effective livestock management in the west.[43]

In contrast, in Norway the traditional cooperative *siida* system is being formalised and used as a basis for re-distributing winter pastures. Reindeer herding is usually organised into summer and winter *siidas*. The summer *siida* was formally recognised by the Reindeer Management Act from 2007[44] and is a more formal institution than the winter *siida*; the summer *siida* is required to have a board that facilitates the practical implementation of collaborative activities. Currently, there are plans to formalise the winter *siida*, primarily through establishing fixed *siida* grazing boundaries and user rules.[45] The redistribution can thus be viewed as a step towards increased co-management, as well as an attempt to reinstate power to the traditional *siida* system by giving *siidas* exclusive user rights to geographically delineated winter areas.[46] The *legal* consolidation of *siida* user rights, however, can be seen as a step towards the privatisation of grazing areas.

In summary, while in China the overall aim seems to be to *re-distribute pastures to individual households* (although both group tenure and individual tenure seem to coexist), in Norway there is a *collective re-distribution* of previously common/semi-common winter pastures.

Fragmentation, privatisation and density dependence

Privatisation as a source of fragmentation

Four global trends in rangeland land tenure change have been described: (1) the maintenance or expansion of state ownership and pastoralist use of rangeland; (2) the quasi-privatisation of state land or devolution to local control; (3) the privatisation of commonly used (often state-owned) land; and (4) the maintenance of private ownership and use with some consolidation or collaborative management of private lands.[47] As described in the previous section, rangelands in both China and Norway were owned by the state (or the lineage or clan in eastern parts of Tibet) but where groups/individuals had some form of user rights to designated tracts of land (albeit informal) and where reciprocal access was prevalent, pasture use was flexible. In contrast, the rangelands in both countries are now being quasi-privatised so that individual households or groups have exclusive user rights, thereby limiting flexible pasture use.

Changing land tenure from commons to private can be viewed as beneficial: it might provide nomadic pastoralists with more control over their own lives as well as provide them with a legal basis for claiming and enforcing rights vis-à-vis competing interests.[48] Privatisation, however, is often followed by *fragmentation*: the dissection of landscapes into spatially isolated parts,[49] often through fencing.

To understand the effect of fragmentation we have to consider how resources are distributed in time and space. In general, fragmentation is only a problem if key resources are distributed unevenly in space (or time). If not, all important resources are present in the fragmented patches (Figure 7.1A).

In contrast, if key resources are distributed unevenly – for example some areas have better quality grass than others, water holes utilised by livestock are only present at some places as in Africa, winter pastures differ from summer pastures as in Tibet and Norway – fragmentation represents a problem because it might destroy the connectivity between important resources. Fencing has the potential to break the connectivity between differentially distributed pasture areas. Due to the high altitude on the Tibetan Plateau, the growing season is short. It starts in late April or early May, and ends in mid-September. The winter pastures are thus especially sensitive: the amount of vegetation left by the end of summer must sustain the livestock until next year's growth begins. This results in a pattern where winter areas are 'saved' for grazing during seasons with no vegetation growth.[50] Fencing is a viable option for protecting these important grazing areas – and has in fact been supported by the Chinese Government through subsidies for the costs of buying and erecting them.[51] The problems arise when everyone fences their 'private' summer and winter pastures: since they are located in different areas, moving between them becomes difficult (Figures 7.1B and 7.1C).

The fact that pastoralists have traditionally been mobile seems to indicate that resources are, in general, distributed unevenly in both time and space.[52] It appears that this simple fact has not been considered in any process of privatising rangelands. Instead, the number of livestock per household has provided a guideline for calculating how much area that household would need as its own private grazing area. In other words, there has been no consideration of the quality or quantity of the different grazing land – and when it has, it has favoured the powerful herders, where they have secured access to the best and largest grazing areas through political influence, as seen in Inner Mongolia.[53]

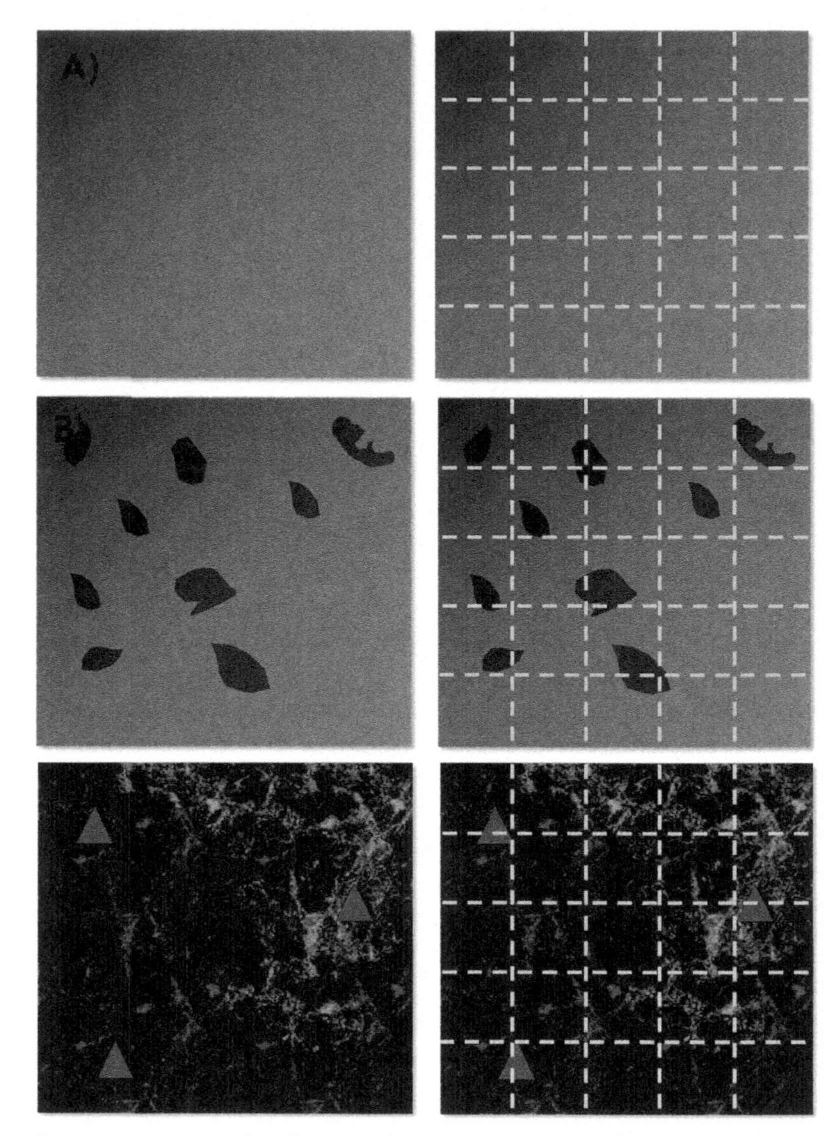

Fig. 7.1 (A) Even distribution of grazing resources – fragmentation by fencing would not be a severe problem as long as the quantity within each patch is sufficient (right panel). (B) Uneven distribution of grazing resources where darker patches represent poor grazing resources. Fragmentation by fencing would represent a problem depending on which patch you occupy (right panel). (C) Uneven distribution of grazing resources and water points (triangles) in time and space. Left corner with darker colour represents summer grazing while right corner with lighter colour represents winter. Fragmentation by fencing would represent a severe problem as herders would have to cross neighbouring patches – owned by other herders – to travel from winter to summer pastures as well as when accessing water points (right panel) (Source: author).

Density dependence and density independence

From an ecological point of view, it is often argued that populations are *regulated* by density-dependent factors (competition, predators, stress, parasites, etc.) and *limited* by density-independent factors (climate, temperature, light, latitude, etc.). The overstocking paradigm takes as its starting point the primacy of density dependence: livestock and pastures are regulated by grazing pressure alone. In contrast, in the early 1990s range ecologists and anthropologists started to argue that livestock and pastures are limited by external factors such as climate (density independency), especially in arid and semi-arid areas.[54]

In systems characterised by density dependence, sustainable levels of grazing are relatively easy to calculate: it can be defined as a relationship between vegetation and livestock. Negative livestock or vegetation growth is seen as a symptom of overgrazing. This is usually conceptualised as carrying capacity: the basic idea being that as livestock numbers increase, available food decreases, which over time negatively affects livestock numbers. The trick is to keep livestock numbers at a stable level – through harvest – creating a balance between numbers and available food.

The problem, however, is that no system is as simple as this: climatic factors like snow or drought negatively affect vegetation irrespective of livestock numbers. In other words, carrying capacity might vary depending on climate. Pertinently, there are also indications that density-dependent and independent effects interact negatively: it has been shown that population growth rates or survival vary more at high density, for example density-independent effects can be stronger at high densities.[55]

The form of density dependence of interest here relates to food availability: as the number of animals increases, competition for food also increases. In general, with more animals, less food is available per individual animal. With less food available, body mass decreases; this is important because there is a positive association between body mass, survival and reproduction.[56] Livestock with poor nutritional status are also more susceptible to disease.[57] Livestock usually gain body mass during the good season (e.g. summer) in order to survive the lean season (e.g. winter): in reindeer husbandry in Finnmark, Norway, for example, there has been a decreasing trend in reindeer body mass[58] and, in 2010, there was a news report that reindeer were starving to death on their way to winter pastures.[59] According to the report, large herds of reindeer moving to winter pastures trampled the vegetation, leaving little food available to subsequent migrating herds.[60] The obvious paradox is that at this time the reindeer should be in good condition having gained body mass during summer. Previously, starvation was mainly seen during a harsh spring or early summer[61] when the reindeer were in poor condition having lost body mass during the winter season.

This form of density dependence does not necessarily indicate increasing numbers of animals – it might also be caused by animals staying too long in a given grazing area, as this does not allow the pastures time to recuperate. Traditionally, both forms of 'overuse' have been offset by moving and changing grazing areas at regular intervals.

Discussion

> Land privatisation creates a paradox for pastoralists: They need both flexible and secure access to land to ensure future grazing, but if they settle on that land to secure it, their lack of movement means poorer livestock production. Often settlement by one family denies other community members access to common resources and interferes with traditionally coordinated grazing systems, especially in times of scarcity (p. 226).[62]

Reduced mobility, intensification and degradation

Mobility has been described as a rational response to seasonal environmental variation.[63] This is fairly obvious when considering large-scale phenomena such as the location of grazing areas. Consider, for example, the migratory pattern of reindeer herders in Norway where some herds move up to ~170 km from winter pastures on the interior to summer pastures along the coast.

Mobility can be classified according to the spatial extent of movement. The seasonal migratory patterns of reindeer and herders are influenced by both climate and geography: for reindeer, the most important diet during the winter is ground lichens which are commonly distributed in relatively dry continental areas.[64] Similarly, as indicated earlier in the chapter, Tibetan herders set aside grazing areas that are only utilised during winter. In other words, the migratory pattern between summer and winter pastures meets the different seasonal needs of livestock;[65] a form of mobility often termed *resource exploitation mobility*.[66]

On a smaller scale, there is *escape-* or *micro-mobility*: movement in order to escape environmental hazards.[67] Tibetan nomads move their herds quite frequently within different seasonal grazing areas, and sometimes even cross into another seasonal grazing area if necessary. Heavy snow during the summer, for example, causes problems: since sheep and goats are poor diggers, the nomads have to wait to bring the sheep and goats out to graze until after the snow has melted. Nevertheless, since it can snow continuously for days on end, it may be impossible to take the animals to the summer pasture. As a consequence, nomads often have to utilise areas reserved for winter grazing during the summer. These winter areas are further from the mountains and thus relatively free from snow during the summer. The ability to move is thus not only restricted to seasonal utilisation of different grazing areas, but also incorporates the ability to respond flexibly to day-to-day variation in climatic factors such as snow.[68]

Mobility in the face of environmental risks has been argued to undergird the survival of most nomadic pastoralists[69] and for centuries pastoral mobility has provided herders with the flexibility needed to survive in patchy, unpredictable and low-productivity environments.[70] Little et al.[71] argue that mobility is the key pastoral risk management strategy; pastoralists who migrate with their herds have considerably fewer livestock losses during climatic disasters than their sedentary counterparts. More to the point, mobility allows pastoralists to take advantage of resources found in different habitat types and thus supports more animals than would be possible if they were stationary.[72]

Pastoral movement therefore seems to be a rational strategy aimed at dealing with the vagaries of the herding lifestyle. Nevertheless, the same strategy has been considered unsustainable and non-rational by national governments all over the world.[73] In fact, privatisation has been implemented as a countermeasure to what has been considered an unsustainable resource use: the assumption being that open access of privately owned livestock to common rangeland has led to severe rangeland degradation. In short, privatisation is assumed to be an efficient tool to combat rangeland degradation.

In contrast, it has been noted in Africa that areas with concentrated use are marked by severe and spreading degradation of vegetation and soils, leading to lower herd productivity and increased herd size requirements to meet household needs. In turn, this accelerates environmental degradation and the probability of poverty.[74] Crucially, privatisation and fragmentation have resulted in an increased concentration of both people and livestock in small areas leading to increased grazing intensification and consequent rangeland degradation.[75]

In Maqu County (eastern part of the Qinghai-Tibetan Plateau) two grassland management patterns currently exist: (1) a traditional multi-household system where grassland is jointly managed by two or more households with no fences between individual households and (2) a single-household system where grassland is separately managed by one individual household and is fenced. A study comparing the respective benefits of the two management patterns found that multi-households were more mobile and that the single-household pattern was more likely to cause rangeland degradation.[76] A study looking at rangeland conditions over time found that while there was no significant difference in 2009, by 2011 multi-household grasslands had significantly higher biomass, vegetation cover and species richness than single-household grasslands.[77]

One study in Inner Mongolia – an area experiencing high level of degradation since the 1980s – reported that 'it is reasonable to assume that the property rights regime change [i.e. privatisation] might be one of the reasons for grassland degradation' (p. 465)[78] and may in fact have accelerated degradation.[79] The same has also been argued for Kyrgyzstan where the '[p]rivatisation of livestock and decreased mobility of herders has in turn led to increased use of pastures immediately around villages, resulting in extensive pasture damage, proliferation of unpalatable woody plant species and large slope failures in these areas' (p. 193).[80] A study comparing changes experienced by pastoral societies and their environments in Mongolia, Inner Mongolia, Xinjiang, Buryatia, Chita and Tuva, found that the highest levels of rangeland 'degradation was reported in districts with the lowest livestock mobility; in general, mobility indices were a better guide to reported degradation levels than were densities of livestock' (p. 1148).[81] In short, due to fragmentation and subsequent reduced mobility, privatisation has been found to exacerbate the same effects it was introduced to counter; the underlying reason being that fragmentation increases density dependence.

The erosion of cooperative networks

The *siida* and *ru skor* systems were small cooperative networks, based on kinship, that flexibly formed and reformed according to both external (e.g. pasture) and internal (e.g. population growth) factors.[82] The *siida* and *ru skor* were cooperative groups based on close kinship ties allowing members to: (1) maintain face to face communication; (2) monitor each other; and (3) punish individuals who broke the rules. These are all characteristics that to a large degree favour cooperation and deter free riding tactics.[83] The *siida* and *ru skor* were fluid and dynamic, their composition could change as a result of expulsion, or alternatively some households left the group and changed partners because of a transgression of rules connected to, for example, the sharing and exchange of labour.[84] Moreover, they have been described as changing according to season: the *siidas*, for example, were smallest during spring calving and largest during the summer.[85]

The inherent seasonality of cooperative group formation was also present among Tibetan herders: since environmental, demographic, political and social conditions vary during different seasons and at different locations, the *ru skor* also changed in size over the course of a year.[86] The importance of cooperative production has been demonstrated theoretically[87] as well as empirically among reindeer herders in Norway, indicating that pastoralists with extensive cooperative networks do better than pastoralists with less extensive networks.[88]

Privatisation and fragmentation may not only break resource connectivity, but also *social connectivity* by dismantling the traditional cooperative networks. As indicated earlier, the *ru skor* seem to have been destroyed – or at least have diminished in importance – on the eastern parts of the Qinghai-Tibetan Plateau. In general it has been argued that privatisation may break up already existing group organisation and prevent 'effective cooperation in herd and rangeland management within and among pastoral communities' (pp. 141–2).[89]

From a general point of view, mobility – specifically the movement of people – has been found to be an important prerequisite for cooperation. The logic is as follows: imagine that you work together in a group with other herders. Suddenly you discover that some of your fellow herders never contribute to common tasks, for example they stay in the tent rather than helping with herding or during shearing they gladly accept help with their own animals but never help out when other herders shear wool from their animals. Traditionally, you would have been able to change group – it is most likely that you would have had family in another group that you could move to. Not surprisingly, the ability to move or change groups is a deterrent for free-riders: the ability to move away allows would-be cooperators to assort positively as well as limit the rate at which cooperators are exposed to defectors. Known as the 'walk-away' hypothesis,[90] there are strong indications that simply providing the option to move allows cooperation to persist for a long period of time.[91] It is difficult to see how such a flexible system of group formation can be upheld in a system with privatised and/or fenced grazing areas that cut across former cooperative groups. Similarly, if group membership becomes consolidated through the legal system – as is the plan in Norway – transferring to another *siida* might become difficult for individual herders. In short, positive assortment, facilitating cooperation, might be limited with land tenure privatisation.

In addition, it has been claimed that privatisation has resulted in increasing levels of conflict and created the potential for new disputes, because fuzzy boundaries are open for negotiation while fencing in rangelands precludes negotiation.[92] Moreover, privatisation seems to have changed the nature of conflicts: previously conflicts occurred primarily *between groups*, now conflicts occur between *individual (former) group members*[93] and also between family members (usually brothers) and neighbouring households.[94] In short, formerly cooperative relationships may have been transformed into competitive relationships.[95] Privatisation has also resulted in increasing differences between poor and rich herders: For example, in Inner Mongolia in the 1980s those with the means to enclose land did so – effectively a first-use principle for those with most power. This intensified economic exploitation and encouraged more irregular grazing practices.[96] Powerful and rich herders therefore enjoyed a tremendous advantage in the local competition for present and future grassland resources; some have enclosed far more than their allotted share.[97]

Concluding remarks and future prospects

While discussing the significance of *place* in the construction of anthropological theory, Appadurai[98] makes a number of observations relevant for this volume's focus on Arcticness. Appadurai[99] notes that there is a tendency for places to become showcases for specific issues over time and thus might restrict theoretical discussions locally as well as exclude other relevant issues. Appadurai cautions us to ask:

> whether these gatekeeping concepts, these theoretical metonyms, really reflect something significant about the place in question, or whether they reveal a relatively arbitrary imposition of the whims of [anthropological] fashion on particular places (p. 358).[100]

Arcticness as a 'quality of being Arctic' – as Medby writes in the preface of this book – has the potential to become a theoretical construct linked to a specific place, that is, the Arctic, that excludes other lines of inquiry.

It also has an explicit *ontological* connotation: while anthropology has had a long tradition of documenting different ideas of what 'is' and how to 'be', it has always been firmly rooted in the idea of a common humanity shared by all people in all cultures.[101] Currently, however, *the ontological turn* posits a move from different worldviews to different worlds altogether; from reality to realities; from variations of how to be human to emphasising incommensurable differences.[102]

It is therefore important to critically investigate what exactly Arcticness denotes. Do we take it to mean ideas about being in the Arctic, for example Arctic worldviews? Or are we positing the Arctic as an ontological distinct lifeworld where the quality of being Arctic unfolds? If the former, then Arcticness becomes an unnecessary theoretical construct that we do not really need. If the latter, then we might reinstate the Arctic and the people who live there as the significant 'Other', fundamentally different. In other words, Arcticness might become a concept of exotification where we reify what it means to live and be in the Arctic. Because by adding –'ness' to the word Arctic, we seem to point to something qualitatively essential, immutable and unchanging with being (and living) in the Arctic, while in fact – as Medby points out in the preface – the Arctic is undergoing rapid changes on several fronts.

As shown here, a comparative approach is fruitful for understanding challenges facing reindeer herders in the Arctic parts of Norway. It might not tell us much about the 'quality of being Arctic' (or, in fact, it might not tell us anything about 'the quality of being a reindeer herder in the Arctic', which to me makes more sense, since it does not have the connotation of 'being a place'), but comparative evidence indicates that privatisation might result in a corollary of unintended consequences for reindeer herders: (1) reduced mobility and increased degradation; (2) increased conflicts and/or the development of social hierarchies; (3) a negative impact on efficient cooperation.

Concurrent with land tenure changes that reduce pastoralists' ability to respond to environmental variability by moving away from affected areas, environmental variability has increased during the last few decades and is predicted to increase further in the future due to climate change.[103] As for the Arctic and sub-Arctic, scenarios generated by most climate models predict that the climate is likely to become increasingly unstable during the next half century with concomitant increases in the frequency of extreme weather conditions.[104]

A case has been made that pastoralists are in a unique position to tackle climate change due to extensive experience managing environmental variability in marginal areas[105] and it has been argued that the ability to withstand environmental shocks is a *defining* feature of pastoralism.[106] Nevertheless, a case can be made that traditional pastoral risk management may be insufficient for dealing with climate change.[107]

In theory – depending on the spatial scale of extreme weather events – mobility has the potential to provide pastoralists with recourse from the most detrimental effects of climate change because they may be able to move away from the affected areas (and thus increase the herds' recuperative potential). I have already discussed the relationship between survival and body mass: animals in good condition are better equipped to deal with harsh environmental conditions. They might survive for a longer period of time during a drought, for example, than those in poor conditions – basically they have a longer window of time to lose body mass before starving to death. In terms of climate change, where we expect – as well as have observed – that the frequency and duration of extreme events like drought, icing, snowstorms, etc., will increase, keeping animals in good condition seems to be an important strategy.

The apparent paradox is that privatisation and subsequent fragmentation has the exact opposite effect: it increases density-dependent food limitation for animals by either intensifying grazing in a limited area or circumscribing too many animals in a limited area, or both. With fencing restricting movement, pastoralists have inadequate opportunity to offset these effects: it is therefore expected that – on average – body mass and condition decrease, making livestock more susceptible to environmental hazards. It should come as no surprise, then, that it has been argued that it is not climate change by itself that is problematic for pastoralists but rather 'the limitations imposed on pastoral coping and development strategies, especially their ability to move and to access critical resources in different territories' (p. 3).[108] Consequently, it may not be mobility *per se* that fails, but rather mobility in increasingly fragmented landscapes.

Another traditional and efficient strategy utilised by pastoralists to buffer environmental variation is herd accumulation.[109] Among Saami reindeer herders in Norway it has been shown that herders with large herds have comparably larger herds from one year to the next[110] as well as before and after crisis periods.[111] While herd accumulation seems to be an efficient strategy, it is predicated on periods of recuperation when herd growth is possible. In fact, a delay in recuperation after environmental-induced losses has been argued to be one of the main problems of pastoral production.[112] Herd accumulation can thus be expected to work less efficiently, if at all, when the frequency of extreme events increases. Pertinently, cooperation is an integral part of pastoral production and has been found to be prerequisite for efficiently accumulating herd size: pastoralists with extensive cooperative networks seem to do better – measured in terms of herd size – than pastoralists with less extensive networks.[113]

Acknowledgement

Funding for this study was provided by the Research Council of Norway (grant number: 240280).

aurora

with terrifying beauty
the sky alight with flame
shimmering and quivering
the soul shan't be the same

the wispy, haunting tendrils
sear down from up above
remote from their oppression
of life and lore and love

the spirit's chilled by wonder
its ceaseless awe is pain
this ecstasy of torture
a universe insane

(Source: author)

Ilan Kelman

8

Energy justice: A new framework for examining Arcticness in the context of energy infrastructure development

Darren McCauley, Raphael Heffron, Ryan Holmes and Maria Pavlenko

We propose the application of an emerging research agenda in 'energy justice' to consider Arcticness in the context of energy exploration in the Arctic region. We define Arcticness as a *process* (rather than a state of being) of bringing voice to those affected by change in the Arctic. It is important not to objectify Arcticness as this will lead inevitably to exclusion. We should instead *subjectify* in the context of past, present and future changing trajectories – a changing process. We therefore need frameworks for exploring and indeed promoting this changing process of 'Arctic voice'. Energy justice is a framework that is able to contribute to this process.

The context of change in this chapter is not the climate, but rather energy exploration. Almost a third of the world's undiscovered gas and 13 per cent of the world's undiscovered oil may be found there, mostly offshore under less than 500 meters of water.[1] In an age of resource depletion, researchers need to pay greater attention to justice concerns in energy policy. In particular, energy exploration – and the resulting energy infrastructure that is built in the Arctic and across the world as a result of the energy resources being extracted – is a major concern for the world. This is even more important when considering the knowledge the global research community published in 2016 and highlighted: (1) temperatures in the Arctic are running at 20°C higher than normal at this time of year;[2] and (2) because of the high temperatures there will be

19 'tipping points' in the Arctic region that will suffer severe consequences and there will be direct effects felt by many countries around the globe.[3] Energy justice provides a framework for assessing the justice implications – or simply the injustices – of current policy decisions as well as making practical recommendations. In this chapter we identify some key injustices and recommendations with regards to uncovering Arcticness. We finish with a call for research into 'frames of injustice' beyond those currently promoted by existing energy justice scholarship.

The energy justice framework

A wide range of the modern-day justice conceptualisations that exist, including environmental, (anti-)global, climate and now energy justice are, to different extents, rooted in finding voice for the excluded. 'Environmental justice' aims to act '(where) people of colour and lower socio-economic status are disproportionately affected by pollution, the siting of toxic waste dumps, and other Locally Unwanted Land Uses (LULUs)'.[4] This has been more successfully utilised as a mobilisation tool for activists in the USA,[5] with some notable exceptions with regards to the protection of indigenous peoples across the Americas[6] or in Taiwan[7] or tribal groups facing environmental hazards in Africa.[8] Through initial explorations of distributive and subequently procedural justice concerns, environmental justice scholars have 'examined multiple reasons for the construction of *injustice*',[9] including race,[10] gender[11] or culture.[12]

'Global Justice',[13] and its more recent incarnation, 'climate justice',[14] emerged from 'anti-globalisation protests', aimed in the first instance at global trade imbalances and then at international climate negotiations. Global justice retains a distinctly economic focus in arguing for the redistribution of existing wealth and indeed new distributions of wealth. Its procedural dimension concentrates specifically on reforming international governance structures. Global and climate justice share, moreover, a common preoccupation with increased recognition of under-represented cultures.[15] Climate justice has, nonetheless, developed a more sophisticated research agenda through assessments of city and locally-based incarnations,[16] in addition to international-level action.

Energy justice (the focus here) carries the same Rawlsian liberalism approach, while incorporating Fraser's recognition of justice and cosmopolitan justice. Two critical distinctions are evident within this research agenda. The concept is, first, rooted to energy systems. In this way, therefore, it aims to provide all individuals, across all areas, with safe, affordable and sustainable energy. We increasingly need a more nuanced understanding of social justice concerns within energy systems, from production to consumption. Energy justice offers, second, a unique opportunity to engage with established thought in science, policy and activism. We will now cover in more detail two core themes or tenets of energy justice that have emerged in the justice literature for energy policy: *recognition* and *procedural* justice.

The framework

Our energy justice framework is underpinned by the principles of cosmopolitan justice. Cosmopolitan philosophy is the belief in that we are all 'world citizens'.[17] With the advent of clear and visible effects of climate change, the approach to environmental protection is being seen more in the light of cosmopolitan philosophy. Cosmopolitanism has, of course, a distinct and long history in global justice thinking. From this perspective, we build on environmental and climate justice demands for a collective approach to resources. The focus here, however, is targeted on energy resources in the Arctic regions in an attempt to achieve a meaningful global change, specifically in terms of energy behaviours and attitudes.

From this perspective we identify two frames of analysis for this chapter: procedure and recognition. An adoption of recognition justice could shed light on under-recognised sections of society. There is often not only a failure to recognise but also to misrecognise and therefore distort people's views, whcih can be demeaning or contemptible.[18] Thus recognition justice includes calls to recognise the divergent perspectives rooted in social, cultural, ethnic, racial and gender differences.[19] Second, energy justice requires the use of equitable procedures that engage all stakeholders in a non-discriminatory way.[20] It states that all stakeholders in the Arctic should be able to participate in decision making, and that their contributions should be taken seriously throughout. It also requires participation, impartiality and full information disclosure by government and industry,[21] and the use of appropriate and sympathetic engagement mechanisms.[22] In addition, due process is relevant to every level of energy decision making at local, provincial, national and global levels. We expand this principle below to consider also the role of the 'non-human'.

Energy infrastructure development in the Arctic

The energy context in the Arctic is dominated by oil and gas reserves and the increasing role of international companies. Extraction and production takes place on the basis of resource ownership. The Arctic states are Canada, Denmark (with Greenland, an autonomous Danish dependent territory, and the Faroe Islands), Finland, Iceland, Norway, Russia, Sweden and the United States. However, according to the 1982 United Nations' Convention on the Law of the Sea (UNCLOS), the right to explore natural resources in the ocean belongs to the coastal states within the distance of their Exclusive Economic Zones (EEZ), that is, 200 nautical miles. Therefore, only six of the Arctic states can legally exploit oil and gas within the Arctic circle, namely Canada, Denmark, Iceland, Norway, Russia and the USA.

Non-Arctic states such as China, Japan, India and Singapore as well as the European Union have expressed their interest in engaging in Arctic-related activities ranging from research programmes to direct extractive operations. Some non-Arctic-based companies take part in joint projects with companies from the Arctic states, for example the Italian company ENI currently has a joint exploration agreement with the Russian organisation Rosneft. This creates a unique operational environment where a few actors representing countries with diverse economic, political and cultural backgrounds are responsible for a vulnerable and complex environment and the intimately linked futures of 400,000 indigenous peoples. The activities of energy companies that are exploring oil and gas in the Arctic are likely to determine the Arctic's economic, social and environmental well-being in the years to come.

Yet, Arctic development is a risky and costly venture. The major drawbacks include the remoteness and harsh climate conditions, which require more advanced technologies, equipment and infrastructure, as well as competition from unconventional gas sources such as shale gas and liquefied natural gas. In addition, there is a long investment cycle and potential overlap of sovereignty claims. The development of Arctic reserves, however, may have serious implications not only for an oil and gas company's budget, but for the global climate in general. Interventions in the fragile Arctic environment may put the future of the region and the planet under a great threat. While the rising demand for resources pushes companies to play for high stakes, environmentalists warn that the consequences of their actions may be irreversible.

Justice and Arcticness in energy infrastructure development

The first tenet of the framework manifests as a call for equitable procedures that engage all Arctic stakeholders in a non-discriminatory way. Arcticness is therefore dependent on voices being heard. Indigenous and non-indigenous peoples are central, for example, to monitoring the increase in tourism in the high north, but equally the intentions of business to develop there. Cultural pluralism is a place for creative industry. Fishing- or reindeer-based livelihoods should be respected. But more attention should be paid to the knowledge creation this involves with its implications for siting and procedural-based decisions. Land use change is a key challenge for indigenous peoples – who moderates if and where land is used for other uses? Holistic management plans are needed which focus equally on the land and not just the sea.

Early intervention is paramount to an effective consultation process. More positive examples were raised also, where companies took a more proactive and constructive approach. As Kadenic concluded in an examination of large-scale Arctic mining projects, the degree of local involvement during the planning phase will directly affect future socioeconomic outcomes.[illegible] From siting decisions to projected habitat destruction, the Saami people, for example, can therefore help developers achieve common outcomes. Procedural justice is more than simply inclusion. It also involves the mobilisation of local knowledge.

A central theme in Arctic energy development is the identification of local communities. Projects in Canada involve multiple indigenous peoples in project development in an explicit attempt to profit from 'multiple views' on local knowledge and creativity. Almost all economic activity in Canada's Arctic is reviewed not just for its economic and environmental aspects but also social factors. However, the involvement of indigenous peoples has been limited. These differing views clearly indicate that a desirable level of economic activity, as well as the extent of being or feeling included in decision making is highly subjective and contextual.

On Russian oil development in the Arctic, there is trilateral policy making: businesses, local governments and indigenous peoples, all of whom need to get their 'fair share' from the activities agreed. Yet the latter group especially are often disadvantaged; for example, they frequently have to endure the low-level jobs which result from development projects. Large corporations come into local communities – where education levels tend to be low – with 500-page technical reports and ask for comments, which is not a fair way to involve the indigenous population. The large size of the corporations involved means that decisions are taken at far away headquarters, while local representatives have to manage their implications for affected communities.

The second tenet of our framework, recognition justice, sheds light on instances of under- or mis-recognition of vulnerability. Local communities such as the indigenous Saami peoples are scattered across most of the northern parts of Norway, Sweden, Finland and Russia, living off fishing and reindeer herding. In addition, there is an under-recognised importance of the non-indigenous people in this area. In both cases, these populations are heavily dependent on local ecosystems. Hence, such communities are extremely vulnerable to energy development.

The richness of fossil fuel energy resources in the Arctic area can be considered in contrast to the provision of energy and electricity in many of those areas. A number of Arctic regions in Alaska are off the electricity grid and electricity has to be generated by diesel generators. This is highly problematic in many ways and contributes (next to health issues) to comparatively low living standards. Such lower standards of living in areas of fuel richness point to local communities having an insufficient level of participation and an inadequate stake in the wealth generated by exploitation activities. As Parlee notes, indigenous communities often have limited access to certain forms of capital and are therefore particularly susceptible to the resource curse phenomenon.[24]

Increasing living standards in the Arctic region is a central mechanism for reducing vulnerability, while simultaneously threatening the environment. The low population density within the Arctic hints at the vast natural space, precisely what makes the Arctic so unique. Tourism in the Arctic region will increase with a growing global upper middle class which is looking for more authentic and exotic holiday experiences. This comes with its own challenges: for example, little effort is put into preserving reindeer herding as one of the large traditional economic activities. Tourism, if exercised in certain ways and at certain scales, will itself contribute to environmental degradation and create issues of a different nature, depriving the Arctic of its unique vastness. Stewart and colleagues report that while the opportunity to educate visitors appears as a positive benefit reflected in the perspectives of residents about cruise tourism in Nunavut, there are emerging risks at the community level which highlight the need for appropriate policies to mitigate the vulnerability of those communities.[25] Therefore, greater involvement of local populations and attention to their knowledge of the region is needed to direct touristic flows. This allows the generation of additional income by offering authentic experiences, while preserving local ecosystems and habitat.

In this context, it is important to consider how extractive industries and other activities potentially impact upon the means of action of local peoples. One dimension is improving general levels of human security. Revenue streams from commercial activities could potentially benefit the security aspect of freedom from want – the provision of an adequate standard of living. In fulfilling this approach, we need to fully appreciate that indigenous groups significantly differ in their histories, and thus in their present needs as well as their visions for the future. Therefore, it is important that different local groups are considered individually within their contexts rather than being seen as all coming from the Arctic region. Thus, the mere engagement of the Arctic community into planning and decision making as an attempt for procedural justice is insufficient. Regional differences across Arctic communities must be respected and taken into consideration.

Beyond indigenous peoples, academic scholars can equally be identified as under- or mis-recognised. A call for the recognition of northern scholars in the identification of research priorities in Arctic areas is also needed. The focus has to be redirected towards the co-production and co-communication of research results between science and stakeholders. Next to a better integration of natural and social science in the Arctic, advancing recognition-based justice would be achieved if research results were presented in a way which is easy for non-scientific audiences to understand. Part of recognition justice is the informed self-determination of future development pathways that communities choose for themselves, despite adherence to traditional social and economic activities.

Expanding justice in Arcticness – a new role for the non-human

One particular debate on Arcticness deserves particular attention in this study, namely whether the natural environment can be considered a separate voice. The energy justice framework continues to suffer from a uniquely anthropological outlook. Arctic-based ecosystems and habitats are at the forefront of energy developments in the region. If their full implications are to be considered, energy justice must be more than a means to 'provid[ing] all individuals, across all areas, with safe, affordable and sustainable energy'.[26] Protection of the environment should have equal status. One avenue suggests that changing reporting procedures for companies, as the primary agent in a largely unregulated area, may provide some modest hope.

Procedural justice refers largely to human populations, with an overconcentration on impacts upon local communities. We of course agree with Marshall and Brown that 'the question of whether to report on the environment is no longer an issue'.[27] But rather than reporting to stakeholders on environmental impacts, we question here whether the environment itself should be considered to be a stakeholder. It is essential that we find new ways to bring the environment into this debate on justice and security in Arctic energy development.

The main controversy in relation to the environment is connected with its non-human nature. Indeed, the environment cannot physically *engage in dialogue* with developers or articulate its interests and concerns. However, there is no denying that the environment is affected by organisational activities, and the organisation likewise can be affected by the environment. This is particularly relevant to Arctic oil and gas companies as resource extraction can cause extreme environmental damage, for example oil spills from an operational accident, and can easily be disrupted by the extreme weather conditions which are typical of this region.

The definition of a stakeholder, namely '*any* group or individual who can affect or is affected by the achievement of the organisation's objectives',[28] does not explicitly specify whether stakeholding is only applicable to people. Technically, there is no reason not to consider the natural environment as a stakeholder just because it cannot speak. Starik compares the non-human environment to the groups that were historically discriminated against and hence deprived of a political voice: slaves, indigenous minorities, the homeless and political prisoners.[29] He argues that, despite not having such a voice, these groups would still be considered as stakeholders, so why should the environment not also receive stakeholder status? The question remains as to what the practical implications of such recognition could be.

The environment can also be viewed as a stakeholder due to its importance to the interests of future generations with regards to *both* human and non-humans. This argument is of particular relevance to the Arcticness debate as oil and gas extraction in this region is likely to increase the speed of the already melting Arctic ice, which will affect the ecological balance by accelerating the process of global warming. Social scientists need to engage with natural scientists in order to theorise how energy developments can be just to both human and non-human.

Implications: energy justice and 'frames' of Arcticness

Injustice – rather than justice – should be the focal point for energy justice research through a more explicit assessment of master frames of 'injustice' in the pursuit of understanding Arcticness. Master frames are collective action frames of Arctic stakeholders that have expanded in scope and influence. Put simply, a master frame encompasses the contextual boundaries, interaction and normative claims of more than one organisation, one movement or one voice. Such frames can indeed vary dramatically in terms of restrictiveness or exclusion. Gerhard and Rucht found that two distinct master frames (with different protagonists, antagonists, organisations, etc.) worked together to encourage social mobilisation in Germany.[30] They can, therefore, often serve as a 'kind of master algorithm that colours and constrains the orientations and activities of other movements'.[31] Scholarship in energy justice research remains theoretically, conceptually and contextually bound. This section concludes with a reflection not only on unbinding energy justice research from pre-set notions of justice, but also its conceptualisation of 'environment'.

Theoretical accounts of energy justice threaten, first, to bind researchers into pre-determined logics of justice.[32] For Caney, justice research has hitherto focused on exposing and proposing archetypal normative frameworks.[33] In support of Agyeman and colleagues,[34] Reed and George comment, 'researchers are cautioned that the long-observed disconnect between theory and practice in the field of environmental justice may be exacerbated should academics become more concerned with theoretical refinement over progressive, practical, and possible change'.[35] The theorisation of justice seeks to expose ideal end points (and more recently processes) from various philosophical traditions. For example, Okereke finds that any notions or principles of justice originate from five distinct incarnations: utilitarianism, communitarianism, liberal equality, justice as meeting needs and libertarianism[36] – later refined to include 'market justice'.[37] In a similar vein, Schlosberg argues that justice theorists need to be pluralist in accepting a range of understandings of 'good'.[38] It is argued here that we need instead to explore the plurality of injustice.

The first step in this direction is indeed the acknowledgement that the study of justice is pluralist. Martin et al. acknowledge, 'that justice poses considerable conceptual challenges, not least because of the practical (if not intellectual) impossibility of reaching consensus'.[39] This is borne out by a valiant theoretical sortie through the myriad of approaches to conclude that justice is both plural and multi-dimensional. Their conclusion bears a self-reflective unease; 'we clearly have much to learn about the limitations of our own framing and methods, including our inevitable starting point in logics of justice'.[40] The second move involves an acknowledgement that justice is contextualist, whereby some principles may apply in certain situations. Walker comments, 'as we move from concern to concern and from context to context, we can expect shifts in both the spatial relations that are seen to be significant and in the nature of justice claims being made'.[41]

Ideal justice theorists seek to effectively eliminate the potential for conflict. Schlosberg comments, however, 'such theorists are mistaken ... (c)onflicts of justice arise ... problem solving entails the negotiation of different conceptions of (in)justice in and across participants, from community or stakeholder groups to corporations or states'.[42] Schlosberg claims that the idea of environmental justice has 'examined multiple reasons for the construction of injustice'.[43] This chapter calls, however, for an exploration of the construction of multiple injustices. An expansion in the theorisation of environmental justice as a concept must be answered with a similar response in our understanding of environmental activism. As Barnett comments in support of Sen:[44]

> Rather than thinking of philosophy as a place to visit in order to find idealised models of justice or radically new ontologies, we would do well to notice that there is an identifiable shift among moral and political philosophers towards starting from more worldly, intuitive understandings of injustice, indignation, and harm, and building up from there.[45]

Second, the recent development of normative concepts of justice looms in a similar manner. There is a sense (to some extent correct) that such concepts are worldly, emerging from situated conflict. They are, however, more often emerging from philosophical debate. A set of normative testable assumptions materialise based upon achieving equity and fairness in the distributional, post-distributional – referred to as 'recognition' largely attributed to Nancy Fraser[46] and developed by Schlosberg[47] – and procedural burdens of environmental risk. We of course explore procedural and recognition forms in this chapter.

However, the analytical objective identification of injustice can be blind to the experiential perception of spatial constructs. The more recent attempt to uncover a third form of energy justice tenets as the 'post-distributive justice of recognition' threatens, for example, to unintentionally disrobe those who are unrecognised of any meaningful agency.[48] Even though Fraser firmly identifies social movements as key agents of change,[49] the emphasis is on the call for 'authorities' and 'policy-makers' to recognise under-represented groups – such as in Walker and Day.[50] Framing research emphasises, in contrast, the need to explore such processes among those who are 'under-recognised' in order to gain insight into the success or not in mobilising against injustices. They are often referred to not as 'victims', but rather as 'non-activists', and as posing a new challenge for justice research.

Third, our approach to energy justice remains contextually bound. In this vein, the energy justice 'master' frame is derived from specific empirical contexts – in this case the Arctic. The origins of energy justice research are accepted to be race- and poverty-based campaigns involving multiple organisations and individuals across the USA merging into a veritable energy justice movement – often cited as beginning in Warren County, North Carolina.[51] And thus, the energy justice master frame in the USA is formed around race, class, gender and the environment. Taylor talks explicitly about the 'environmental justice paradigm' as a master frame which links together 'environment, race, class, gender and social justice' issues.[52] In the UK (especially among non-governmental organisations or NGOs), the master frame has been termed as 'just sustainability'[53] despite the earlier observation that there exist 'at least three different constructions of environmental justice'.[54] This refers to a frame that links together issues of sustainability, social inclusion and procedural equity.

Dawson demonstrates, however, the potential fluidity of the energy justice master frame in linking it explicitly to eco-nationalism.[55] She identifies sub-group identity, social justice and environmentalism as the core tenets of the US energy justice frame. The US environmental movement is, in her view, built on the foundation of sub-group identity and the desire for social justice. As a result, groups defined by religion, gender, national identity or class could offer a basis for energy justice movements and their master frame. In this way, the energy justice frame covers, for example, the protection of indigenous peoples across the Americas[56] or Taiwan[57] or tribal groups facing environmental hazards in Africa.[58]

In such a conception, the energy justice frame can actually be ultimately divisive and exacerbate violent conflict. Dawson traces the environmentalist roots of nationalist movements in the former USSR which lead directly to social tensions and fragmentation. She observes, 'the intertwining of environmental causes and sub-group identities can be seen to both enhance environmental mobilisation among previously unmobilised groups and deepen a pre-existing sentiment of "us" versus "them" within the population'.[59]

Empirical conceptions of justice are, therefore, as problematic as theoretical and conceptual incarnations. Pellow and Brulle argue, indeed, that '(s)cholars cannot understand … environmental injustices through a singularly focused framework that emphasises one form of inequality to the exclusion of others'.[60] Our attention should be drawn to where and when injustice is felt and experienced. Hobson argues that energy justice research must diversify its understanding of where injustice can be found. In her assessment of an environmental organisation in Singapore, she demonstrates how environmental injustice is felt in everyday practices of individuals and organisations, even where expressions of public concern on the environment are infrequent or at least highly managed.[61] More recently, substantial research has focused our attention on injustices within climate activism.[62] The fluidity of master frames on energy justice offers one potential solution to unbinding how we approach justice and injustice. We should turn our attention to unlocking further how we can explore master frames of injustice through a better understanding of Arcticness framing.

9

Understanding Arcticness: Comparing resource frontier narratives in the Arctic and East Africa

James Van Alstine and William Davies

Introduction

When exploring what is meant by 'Arcticness' it becomes pertinent to ask what is unique about the Arctic as a 'resource frontier'. The Arctic has found itself receiving greater international attention in recent years,[1] this attention commonly attributed to pronounced sea ice loss from rapid climate change[2] and the subsequent increased accessibility to the region's abundant natural resources.[3] While excitable claims of a region opening up and a resource rush are arguably hyperbolic,[4] nevertheless it is clear the presence of extractive industries will continue to grow in the coming decades, be it offshore petroleum in the Pechora Sea,[5] diamonds in Nunavut,[6] or rare-earth minerals mining in Southern Greenland.[7]

Resource frontier narratives represent a specific set of ideas and interests and can be conceptualised as relational spaces where economy, nature and society co-construct.[8] Typically host governments, international financial institutions and the private sector use this rhetoric to legitimise foreign direct investment, natural resource extraction and commodity production.[9] To investors, the term 'frontier' denotes higher levels of risk but also the possibility of significant rewards. These so-called frontiers are typically located in remote regions lacking strong forms of state governance. The potential exists for higher levels of political, social, technical and environmental risk from resource extraction.[10]

When exploration begins and the first commercially viable energy or non-energy minerals are discovered, the idea of resource wealth leads to multiple imaginaries. On the one hand, the resource curse narrative demonstrates the links between natural resource wealth and weak development outcomes.[11] On the other hand, the resource-led development narrative highlights how new-found petroleum resources may catalyse national development towards middle- or high-income country status.[12] It is assumed that if social and environmental issues are well managed then the extractive industries can contribute to sustainable development and poverty reduction. Indeed, the resource curse narrative has been reframed as a 'governance issue' and a political-institutional challenge, as opposed to a quasi-automatic phenomenon that resource-rich countries are destined to follow.[13]

Predicted environmental transformation and transition towards a resource frontier presents significant challenges for the Arctic[14] as it does for other regions around the world. Here, a comparison with another region experiencing the rhetoric of extractives-led growth proves useful. Significant reserves of oil and gas have been discovered in East Africa in the last decade, prompting governments in the region to model pathways towards middle-income status with significant emphasis on resource-led development. Where the overriding political economic context in the Arctic and East Africa is that of extractives-led growth, it becomes pertinent to explore the similarities and differences between these emerging resource frontiers.

This chapter compares and contrasts developments in the Arctic and East Africa by examining key material, global interest, governance and community themes associated with increased oil exploration in Greenland and Uganda. Both contexts are considered to be at the 'frontier' of extraction given their 'unconventional' locations.[15] In doing so, it aims to better understand the characteristics associated with regions undergoing extractive-led growth imaginaries as well as unpacking location-specific idiosyncrasies such as 'Arcticness'.

Background

Greenland is the world's largest island (2,150,000 km^2) with the vast majority of this territory comprised of 80 per cent ice cover.[16] With a population of 58,000, it is also one of the least-densely populated territories in the world with 15,000 of its population found in the capital Nuuk to the south-west of the country. A Danish colony for over 200 years (1721–1953) with a demographic that is 90 per cent Inuit and 10 per cent 'European', in 1979 was granted 'autonomous rule' within the Kingdom of Denmark, and has since been progressing towards complete independence;[17] what has been described as a process of decolonisation.[18] If this happens, Greenland would become the first Inuit nation state.[19] Further autonomy was granted to Greenland in 2008 under the status of 'Self-Rule' which significantly gave the country control over its vast natural resource reserves[20] which include gold, diamonds, iron ore, cryolite, lead, zinc, molybdenum, oil, natural gas, uranium and other rare-earth minerals.[21] Uranium reserves are considerable near the site of Kvanefjeld with some predicting Greenland could potentially overtake China as the world's largest exporter of the mineral.[22] While some extractive activity has previously occurred in Greenland, no such activity has taken place in recent decades.[23]

Uganda is a relatively small landlocked country (238,000 km^2) in East Africa. It has a diverse landscape with mountains and lakes ringing a plateau. The country is generally tropical (although semi-arid in the north-east) with two dry seasons punctuating rainy weather. It is a densely populated country with a population of over 37 million. As a presidential republic, Uganda is a sovereign state that became independent from the UK in 1962. However, the country was besieged with conflict under dictatorial regimes until President Museveni's regime, which began in 1986, brought relative peace, stability and economic growth to the country. Although a leader in implementing neoliberal reforms in the 1990s, which paved the way for an era of economic growth and positive donor relations, government–donor relations have deteriorated over the last decade with the government hampered by allegations of widespread corruption.[24] While poverty has declined in Uganda from 56.4 per cent to 19.5 per cent between 1992 and 2012,[25] it still remains very low on the Human Development Index (163rd out of 188 in 2016). Agriculture is Uganda's most important sector and employs over a third of the workforce, with coffee, tea, cotton and fish accounting for the bulk of export revenues.[26] Uganda's key resources include copper, cobalt, hydropower, limestone, salt, phosphate and now oil.[27] However, given relatively small deposits, the extraction of minerals has been a very small proportion of export revenues in the past.

Materiality

Within this chapter, materiality is concerned with oil's physical and economic properties, as well as the social-technical and environmental implications of oil extraction in resource frontiers. In Greenland, hydrocarbons have largely been explored offshore in the Disko Bay area, whereas in Uganda this process is being undertaken onshore in the remote Albertine Graben region of Western Uganda. Both regions are remote, difficult to access and pose socio-technical and environmental challenges. However geography matters, there are a number of 'development traps' that hinder income growth in poor countries.[28] One of those 'traps' is being landlocked with poor infrastructure connections to the sea through neighbouring countries, as is the case with Uganda. Greenland, on the other hand, is well-positioned to develop and exploit its resources through seaborne trade. Nevertheless, the risks are significant. In Disko Bay the environmental risks of oil extraction include sea ice, extreme cold and harsh weather. In both areas oil extraction may disrupt local livelihoods, wildlife and tourism given the extensive infrastructure that needs to be built in and around villages, towns and national parks.

Both contexts lack oil infrastructure. Oil reserves in Uganda are extremely remote and isolated from markets, as they are in Greenland. However, the type of infrastructure needed to extract the onshore oil in Uganda is significantly different from Greenland's offshore oil. A 1400 km oil export pipeline is currently being scoped from Uganda to the Indian Ocean via Tanzania, as is a small oil refinery in Uganda for domestic and regional consumption. In Greenland, considerable infrastructure developments to areas such as port and onshore support facilities are required to make oil production and exportation a reality.[29] Furthermore, there are questions surrounding the efficacy and reliability of current oil drilling technology and its ability to withstand extreme polar conditions.[30]

Commercial viability is also an important material attribute of these frontiers. In Greenlandic waters the quantities are estimated as vast (although no commercially viable wells have been drilled yet), with nearly a fifth of undiscovered oil resources in the Arctic region located in two Greenlandic provinces: East Greenland Rift Basins and West Greenland–East Canada.[31] In Uganda, commercially viable oil was discovered in 2006 with recoverable reserves estimated to be at least 1.4 billion barrels of crude with proven reserves of 6 billion barrels. However, given these technical challenges and associated political and economic concerns, progress towards commercial production has been slow in Uganda with the first oil projected for 2020. Indeed, US$50/barrel is needed to make oil production economically feasible in Uganda. For Greenland, the per barrel cost required for economic viability is significantly higher given the expensive production costs involved with working in remote, polar waters and a limited drilling window confined to the summer months.[32] With the crude oil prices dropping as low as US$30/barrel in January 2016, the prospects for offshore oil development in Greenland look increasingly unlikely in the short-term.[33]

Broadly-speaking, there are certainly similarities between the material challenges facing each resource frontier; for example, limited oil infrastructure, economic viability, technological challenges and the potential for socio-environmental impacts. These similarities are, however, unsurprising as such characteristics are usually common to resource frontiers.[34] The very fact that they are frontiers signals limited infrastructure in relation to the resource being developed. Questions over economic viability are always present in such contexts: if it were not so, it is likely the resource would have been developed earlier. Technological challenges are often an obstacle that has previously prevented resource development; likewise are the potential risks associated with negative socio-environmental impacts.

One aspect of Arcticness then is the *extremity* of these material challenges for Greenland. The sparse population of Greenland, the remoteness of its seas, the harsh, polar environment and the vulnerability of ecosystems that are crucial for Arctic peoples' livelihoods are all factors contributing to this extremity. These are, also, characteristics commonly associated with traditionally romanticised imaginaries of Arcticness: a remote wilderness to be explored that challenges human endeavour (and its technological ingenuity) to the utmost.[35] Much in common with the Arctic explorers of the past, there is an aspect of discussing oil company exploration in Greenland that sounds like explorers entering a remote, challenging corner of the globe. On the other hand, geologists have known about the presence of oil in Uganda's Albertine Graben since the 1920s, with the first exploration well drilled in 1938.[36] It was not until the commodity super cycle of the 2000s and the rise in oil prices that exploration into the region was reinvigorated and commercial quantities of oil discovered. Thus a key distinction between Uganda and Greenland is the extremity of oil development in Arctic waters.

Global interest

The Arctic is a region of significant global interest. It is often framed as a region opening up to stakeholders worldwide and home to a new geography of voices.[37] As such, what takes place in the Arctic is of increasing global concern and can often lead to the Arctic resembling a 'global commons', especially given the context of its ocean and climate change. The East African context has interesting parallels.

Two prominent reasons contribute to the perception of the Arctic Ocean as a global commons. First, there is a conflation with the situation found in the Arctic's polar relation in the south, Antarctica. While sharing many cryospheric characteristics due to their polar latitudes, there are certain factors that ultimately define them as fundamentally different international spaces.[38] Crucially, Antarctica is an uninhabited landmass surrounded by ocean. This lends itself more readily to the perception of an international space or global commons. While it is too simplistic to label Antarctica purely as an 'international space', international agreements such as the Antarctica Treaty embody this perception.[39] In contrast, the Arctic is an ocean surrounded by inhabited landmasses. However, in terms of remoteness and lack of inhabitants, the (varying) ice cover at its ocean's centre resembles a polar wilderness akin to Antarctica. It is often these imaginaries of the upper Arctic Ocean that are evoked when the Arctic Ocean is described as a 'global commons'.[40] The ambiguity of the Arctic Ocean[41] facilitates this global perception of the region.

In a Greenlandic context, where oil and gas development is offshore but takes place in sovereign waters, there is a blurring with the high seas of the Arctic Ocean that accentuates a greater legitimisation of 'outsider concern'; for example globally-focused environmental campaigns such as Greenpeace's 'Save the Arctic' around Arctic offshore oil and gas.[42] While the African context is obviously very different, there is still significant global interest in the region. On the one hand, there is often heavy involvement by donors and international NGOs among others keen to assist least developed country governments, such as Uganda, to utilise natural resource rents for pro-poor development (as will be discussed below in the governance section). On the other hand, the general public and media in Western countries make the mistake of homogenising the African continent as one country in dire need of charity and overseas development assistance based upon negative post-colonial narratives of war, disease, poverty, corruption and starvation.[43]

The Arctic's tightly woven relationship with climate change ensures global interest in Greenland's oil adventure. Much is written about the impact of climate change on the Arctic, where the impacts are felt more keenly than anywhere else.[44] Projections from the IPCC and ACIA suggest a surface air temperature rise of 2.5–7°C by the end of the century; significant changes to the Arctic biome's biodiversity and ecosystems;[45] and a substantial retreat of the summer sea ice extent. Furthermore, these changes have significant global ramifications, with a sea-level rise and alterations to global oceanic circulations from the effect of lower surface albedo and increased freshwater run-off from glaciers and/or ice sheet retreat.[46] The rate at which the Greenland Ice Sheet recedes is particularly important with regards to global sea-level rise.[47] It is in this context that Greenlandic oil development finds itself positioned. That there is a commonly held belief that the potential for oil development in Greenland is a direct consequence of increased accessibility from a warming environment reinforces the association with climate change. As such, Greenland's location in the Arctic, at the frontline of climate change, ensures that the discourse around Greenland oil becomes entwined with the wider global climate change discourse.[48]

The African context is significantly different. The exploration and production of oil is not dependent upon a changing climate, but sub-Saharan Africa is acknowledged to be among the regions which are most vulnerable to climate change.[49] The East African region, in particular, has a high dependence on rain fed agriculture, biomass and rivers for energy, and increasing vulnerability given the higher frequency of extreme weather events such as droughts and floods.[50] One of the hardest hit sectors has been energy given the region's reliance on hydroelectric power.[51] While 80 per cent of Uganda's energy supply comes from large hydro, increasing industrial demand and potential delays to large hydro projects may necessitate the recommissioning of heavy fuel oil power plants to fill predicted energy shortfalls.[52] There is significant inequity in energy access as 85–90 per cent of the country's population has no access to electricity. A significant 90 per cent of the people in Uganda live in rural areas and use biomass (wood and charcoal) as primary energy sources.[53] While a changing climate is significant for both Greenland and Uganda, the narratives of oil exploration, climate change and global commons are tightly interwoven within the context of Arcticness.

Governance

As highlighted above, there is significant global interest in how to govern resource extraction in frontier contexts. Governance in this case can be defined as the hard and soft rules that shape and constrain oil exploration and development.[54] The governance of a resource frontier is characterised by a myriad of actors and ideas. International NGOs, donors, multi-national oil companies, and international finance institutions among others vie for policy space to have their ideas and interests taken up and implemented by host governments and exploration companies. Often international norms on transparency, stakeholder engagement, environmental protection and social welfare are advocated given a lack of state capacity in these remote regions.

At a Greenlandic national level, the Minerals Resources Act of 2009 is the guiding legalisation surrounding oil development.[55] The growth in interest in Greenland's oil reserves and the willingness of successive Greenlandic governments to develop this sector has led to governance changes at a national level. Most notably, after pressure from domestic and international civil society, the Bureau of Minerals and Petroleum, which once had responsibility for every aspect of oil development including environmental matters, has seen many of its responsibilities move to new organisations, such as the Environmental Agency for the Mineral Resources Area.[56] Positioned within the Kingdom of Denmark, Greenland governance structures around offshore oil are complex.[57]

Furthermore, Greenland is positioned within a modern narrative of Arcticness: 'as an apolitical space of regional governance, functional co-operation, and peaceful co-existence'.[58] The Arctic as a global commons provides the basis for a spirit of collaboration and cooperation that is somewhat unique and distinct from other regions of the world. This spirit is embodied by the Arctic Council, a non-regulatory, soft-law intergovernmental forum comprised of the eight Arctic states, indigenous groups in the form of 'permanent participants' and 'observers' that include non-Arctic states and other interested stakeholder groups. The council is continually evolving from its original narrow environmental focus and while changes have not been particularly revolutionary, they are nevertheless significant enough to suggest an attempt at adaption to regional change. It is in recent years that governance initiatives around offshore oil development have shown hints of a future Arctic Council transforming into a decision-making body, such as the binding 'Agreement on Cooperation on Marine Oil Pollution, Preparedness and Response in the Arctic'.[59] If such evolution continues, Arctic regional governance could potentially have a greater impact on Greenland's decisions around oil; the contemporary geopolitics of Arcticness influencing its governance arrangements.[60] However, at present, many argue that agreements like the council's oil spill response are weak and require little from the Arctic states who have signed it.[61]

There are also norms for greater continental integration and cooperation for development in Africa. The African Union was established in 2001 and consists of 54 member countries. It is headquartered in Addis Ababa, Ethiopia and made up of both administrative and political bodies. Utilising the extractive industries for sustainable growth is a strong norm in Africa, underpinned by the African Union's 'Agenda 2063' as well as the New Partnership for African Development (NEPAD) and the African Mining Vision, among other African Union entities and documents.[62] While the African Union does have the ability to make binding decisions, for example, through its Peace and Security Council to impose sanctions and deploy peace-keeping forces, it is likely to use soft rules and norms to facilitate extractives-led growth across the continent.

In East Africa the rhetoric of resource-led development has largely been adopted by countries with significant hydrocarbon deposits. In Uganda expectations of the benefits are extremely high: the Government of Uganda's Vision 2040 to transform Ugandan society 'from a peasant to a modern and prosperous country within 30 years' are largely predicated on revenues from oil and gas.[63] While the integration of oil and gas infrastructure has been discussed at the level of the East African Community, cooperation remains elusive as countries compete aggressively for foreign direct investment and oil infrastructure projects.

For example, Kenya and Tanzania sought to win over Uganda as the preferred pipeline route to ports on the Indian Ocean. The Kenyan route seemed most likely for years, until Uganda was swayed in 2016 to partner with Tanzania due to security, economic and land acquisition concerns along the northern route.[64] Thus governance of oil and gas development at the regional level remains mired in zero sum politics between nation states.[65]

Domestically, Uganda's oil and gas legal framework has been influenced by international norms and standards but there are concerns about loopholes and implementation deficits. For example, Uganda's 2008 National Oil and Gas Policy was influenced by international norms on transparency, such as the Extractive Industries Transparency Initiative (EITI). Although Uganda has not adopted the EITI, the policy conforms to international best practice on stressing the importance of transparency and accountability in all aspects of natural resource management.[66] In order to put the 2008 National Oil and Gas Policy into practice, the Government of Uganda has established its legal framework for petroleum: the Petroleum (Exploration, Development and Production) Act 2013; the Petroleum (Refining, Gas Processing and Conversion Transportation and Storage) Act 2013; and the Public Finance Act 2015.

While Uganda's legal framework is seen as a positive step towards transparency and accountability throughout the value chain of the sector, there are a variety of concerns, including: the centralisation of power in the executive branch of government; loopholes in the Public Finance Act that may fall short of EITI requirements; and the National Environment Management Authority's lack of capacity to monitor and regulate the environmental impacts of oil activities.[67] The promise of future oil revenues will most likely reduce Uganda's reliance upon donor budget support, but may have negative impacts on governance, as one political analyst observed in 2006: 'But of course, depending on how commercial the oil is, his [Museveni's] foreign policy will change. He will no longer need donor money to buy political support'.[68]

Community

As Oran Young notes, romantic notions of the Arctic have traditionally represented 'human life in the Arctic that casts the indigenous peoples of the Circumpolar North as happy hunter/gatherers living a simple existence in harmony with the natural environment and uncorrupted by the forces of modernity'.[69] Intuitively, this romanticised view would suggest a reluctance of many Greenlandic people to embrace oil development as it would clash with their traditional lifestyles, from fears over impacts on whaling migration and oil spillage to the social upheaval involved in large-scale industrialisation. In reality, indigenous lifestyles and identity in the contemporary Arctic are considerably more nuanced. This is especially so for Greenland, which serves as an example of indigenous self-government,[70] a situation rare for many of the world's indigenous populations.

Research exploring local community opinion around oil development in Greenland shows a spectrum of perspectives, ranging from those who strongly support the development, those undecided or uncertain, to those vehemently opposed.[71] Those who side more favourably with development tend to cite the important economic benefits that would arise from such large-scale industrial activity and the positive impacts this would have on employment, educational opportunities and healthcare. Those opposed often point to the environmental damage such activity would inflict, both locally and globally, as well as social concerns such as the democratic implications of a small nation becoming dependent on large oil companies and the societal implications of the immigration of foreign workers.[72]

While issues on how oil development impacts indigenous communities and lifestyles are evident and are of importance,[73] they are not necessarily as pronounced as images of Arcticness might suggest. To take one example, research with inhabitants of the town of Aasiaat, a community that served as the base of oil exploration in 2010, found a complex picture regarding the relationship between traditional livelihoods and oil development.[74] For some, traditional livelihoods no longer really existed or were already disappearing, the modern reality being that the majority of Greenlanders now live in towns. Others pointed towards the ability of traditional culture to adapt to changing societal pressures as a fundamental part of the Inuit identity, that is not a fixed position but is malleable to change.

While community concerns found around oil development in Greenland may resemble other areas of oil development, there are differences. In Uganda, the general public, media and civil society appear to largely support the pro-oil resource-led development rhetoric of the national government, that oil should be used to propel the country into middle-income status over the next 30 years. The national debate is around good governance of the resource as opposed to whether or not it should be extracted. At the local level, communities in the Bunyoro region along Lake Albert are comprised of peasant small holders largely reliant upon subsistence agriculture and fishing. The villages where oil exploration has taken place have had significant interaction with oil companies and their contractors. This interaction has led to some casual labour and corporate social responsibility and social investment projects in the areas of health, water, sanitation, education, road infrastructure and local economic development among others. Expectations of local benefits are high, particularly in the area of local employment, infrastructure development and social amenities. However, there are concerns in Uganda's oil bearing communities about lack of information, stakeholder engagement, local employment, livelihood and environmental impacts among other things.[75] These issues are significant, but are not unique to Uganda.[76]

While elements of these narratives are present in the Greenlandic context, the picture is more complicated. Greenlanders are conflicted over various aspects of oil development. This is highlighted in a poll of 721 Greenlanders undertaken in autumn 2013, where it was found half supported drilling and just under half opposed the activity.[77] Oceans North Canada, who conducted the poll on behalf of the Inuit Circumpolar Council, Greenland, cited Greenlanders' strong cultural connections to the sea combined with the perceived risks this would have on important sectors such as seafood and tourism as reasons for those unsupportive of oil development.[78] However, much like Uganda, questions of good governance are ubiquitous within the debate, with the same poll highlighting how the majority felt the government were not doing enough to inform the public, remain transparent or put adequate safeguards in place.[79]

It is relevant to note that community perceptions of oil development will depend on the stage of the project cycle, proximity to the oil wells and associated infrastructure and cultural and economic ties to the areas of extraction. Much Arctic oil development, as in the case of Greenland, finds itself at an early stage of the project cycle. As such, they are surrounded by considerable uncertainty and many unknowns exacerbated by rapid regional change. In Greenland, where oil development is proposed to take place in seas that represent sites of cultural and economic significance, community opposition, or at least greater scepticism, is likely.

Conclusion

This chapter has sought to better understand the concept of 'Arcticness' through the lens of resource frontier narratives in Greenland and Uganda. Somewhat surprising is the amount of similarity in the discussion of material, public good, governance and community themes. On the material characteristics of oil development, both contexts exhibit limited oil infrastructure, challenging economic conditions, technological hurdles and the potential for significant socio-environmental impacts, while Arcticness is characterised by the extremity of oil development in Greenland. Both the Arctic region and sub-Saharan Africa spark global interest; for example creating a marine sanctuary in the Arctic Ocean's international waters and the global fight against poverty in less developed countries such as Uganda. While climate change, a challenge with significant global interest, is present in both resource frontier narratives, it is dominant in Greenland and thus tightly interwoven with the concept of Arcticness.

There are also similarities related to the challenge of governing the sectors in Greenland and Uganda, including the politics of getting the legislative frameworks in place, potential governance loopholes, and the influence of international actors in the 'good governance' of the sector. Modes of cooperative governance are characteristic of the Arctic and at the pan-African level, while cooperation remains elusive in the context of East African oil development as countries compete aggressively for foreign direct investment and oil infrastructure projects. Thus governance of oil and gas development at the regional level in East Africa remains mired in zero sum politics between nation states. Finally, communities in both contexts felt that a lack of information and transparency about the sector were significant issues. However, in Uganda the public is generally pro-oil whereas in Greenland sentiment is rather more mixed.

This chapter has demonstrated that Arctic resource development actually shares much in common with resource frontiers elsewhere. Nevertheless, Arcticness is still a prominent feature of Greenland's oil story. The extremity of oil development, the impact of climate change, the opportunities for cooperative governance, and the mixed community sentiment on whether oil development should proceed highlight aspects of 'Arcticness' in Greenland's resource frontier.

10
Scopes and limits of 'Arcticness': Arctic livelihoods, marine mammals and the law

Nikolas Sellheim

Introduction

The Arctic has become a prominent feature in the current media landscape. It seems fair to say that not a day goes by on which climate change, perceived geopolitical tensions and resource exploitation do not surface in popular media outlets. As a consequence of the environmental changes altering the narrative of the 'frozen north', the Arctic has moved from being perceived as a frontier to a centrepiece of everyday discourse. But only recently, and most notably with the establishment of the Arctic Council in 1996, the notion of the Arctic as a home has entered the popular understanding on the north.

This chapter explores how the Arctic and its peoples are perceived and constructed in international legal regimes. By focusing on controversial marine mammal hunts, in particular the hunt for seals and whales, it examines how legal regimes construe 'Arcticness',[1] how the parameters are set to determine a legitimate or illegitimate hunt and in how far Arctic economies are consequently framed. A focus on marine mammal hunts, which constitute a highly emotional activity, was chosen as they best exemplify the difficulties of reconciling the *imagination* of Arctic cultures and the *empirical realities* in Arctic communities. While drawing on a hermeneutical analysis of available documentation, the chapter is complemented by the author's observations at the 66th meeting of the International Whaling Commission (IWC) in 2016.

Arctic peoples in international legal regimes – the case of marine mammal hunting

From very early on, the Arctic has played a crucial role in international legal regimes. This was not related to the perception of the Arctic as a distinct region, but rather to the abundance of marine resources – particularly whales and seals. Their overexploitation prompted states to conclude bi- and multi-lateral agreements in order to ensure the sustainability of the stocks and, later on, to sustain a viable industry for their exploitation.[2]

The first international environmental treaty that was concluded was indeed based on the overexploitation of seals in the Bering Sea. Throughout the late nineteenth century, American, British/Canadian, Japanese and Russian vessels significantly overharvested northern fur seals (*Callorhinus ursinus*). With the purchase of Alaska from Russia in 1867, the seal-rich Pribliof Islands became part of the United States, which required licences for the hunting of seals in American waters. In the early 1890s, it seized two British/Canadian vessels which were conducting the extremely wasteful pelagic seal hunt without a licence. The case was brought before the Tribunal of Arbitration in Paris. In 1893, the first bilateral treaty relevant for the Arctic was concluded with the Arbitration Treaty,[3] which established a ban on pelagic whaling and set clear rules for the conduct of seal hunting. Most notably, for the purposes of this chapter, the first reflection of the Arctic as a cultural space occurred in this treaty, the narratives of which, by and large, remain until today: Article 8 establishes an exemption for the native people of the region who are not bound to the provisions of the Arbitration Treaty. This, however, is only the case if they conduct the seal hunt in a traditional manner, using traditional equipment and without any commercial intent.[4] The Arbitration Treaty therefore establishes the first legal definition of 'subsistence hunting' and the conditions for exempting indigenous peoples from any bans on or restrictions to commercial hunting of marine mammals. The provisions of the Arbitration Treaty concerning Arctic peoples and cultures were also mirrored in the Convention between the United States, Great Britain, Russia and Japan for the Preservation and Protection of Fur Seals of 7 July 1911 (Fur Seals Convention). The Fur Seals Convention was concluded due to the continuous decline of the northern fur seal populations, primarily because of ongoing hunts by Japanese and Russian vessels. Article IV of the convention reads:

> It is further agreed that the provisions of this Convention shall not apply to Indians, Ainos, Aleuts, or other aborigines dwelling on the coast of the waters mentioned in Article I, who carry on pelagic sealing in canoes not transported by or used in connection with other vessels, and propelled entirely by oars, paddles, or sails, and manned by not more than five persons each, in the way hitherto practiced and without the use of firearms; provided that such aborigines are not in the employment of other persons or under contract to deliver the skins to any person.

The regime collapsed during World War II, being succeeded by the 1957 Interim Convention on Conservation of North Pacific Fur Seals. Article VII of the interim convention once again almost verbatim reiterates the provisions of its predecessor and exempts 'Indians, Ainos, Aleuts, or Eskimos' from the convention's provisions, provided they are not in the employment of commercial enterprises and conduct the seal hunts with non-modern equipment. Up until 1985, when the fur seal regime collapsed,[5] Arctic indigenous peoples were not entitled to technological development when conducting seal hunting in the Bering Sea, but were forced to apply techniques and utilise technology even though safer and more humane practices and technologies were available.

In a similar manner, although not exclusively limited to indigenous peoples in the Arctic, the international regime regulating whale hunting has incorporated exemptions for indigenous peoples. In particular, the whale hunts in Alaska and Greenland played crucial roles in inserting the category of 'Aboriginal Subsistence Whaling' into the work of the International Whaling Commission in 1981.[6] From the first emergence of a whaling regime in 1931, aboriginal people were excluded from any regulatory means. As with the Fur Seal Regime, this exemption only applies when the hunters do not employ modern technology and are not part of a commercial enterprise. Both the 1931 Regulation of Whaling and the 1937 International Agreement on Whaling underline this approach. The 1946 International Convention for the Regulation of Whaling (ICRW) has, however, stepped away from blocking aboriginal peoples from utilising modern technology. Instead, the narrative of 'subsistence need' has been inserted. While this concept in its inchoate form in the whaling context merely referred to local consumption of gray and right whales by indigenous peoples, over time the concept has evolved. 'Subsistence needs' are not to endanger the population status of a species while they are required to correspond to the nutritional and cultural requirements of the respective indigenous people. In order to determine whether quotas for indigenous peoples are set,[7] national governments are to provide the IWC with a 'Needs statement', which 'details the cultural, subsistence and nutritional aspects of the hunt, products and distribution'.[8]

The reflection of narratives that were applied more than 100 years ago in legal regimes affecting Arctic residents are still relevant. The most prominent example is the EU regime banning the trade in seal products that was adopted in September 2009.[9] Here, Arctic livelihoods relating to the hunting of seals are clearly narrated and, throughout the preparatory process of the regime, the 'traditionality' of seal hunts – meaning the long-standing history of seal hunting and processing by *indigenous peoples* – stands at the fore. The Seal Regime's overall purpose is shady, but it functionally bans all trade in seal products in the European Union with the exemption of those stemming from indigenous peoples. This so-called 'indigenous exemption' is, as can be seen above, a common feature in regimes managing – or banning – the utilisation of marine living resources. In the wake of the challenge of the regime before the World Trade Organization (WTO) by Canada and Norway, the EU was forced to amend the regime in order to make it fully compliant with international trade law and its moral exception under GATT Article XX (a).[10] These amendments saw an insertion of animal welfare requirements into the indigenous exemption, but did not alter the way legitimate Arctic seal hunting is legally constructed in a European context. The text of the amended Basic Regulation in Article 3 ('Conditions for placing on the market') thus stipulates that seal products can be placed on the EU market when: (1) Seal products result from hunts conducted by Inuit or other indigenous communities; (2) There is a tradition of seal hunting in the community; (3) The hunt contributes to subsistence and income support yet without primarily commercial intent; and (4) The hunters pay due regard to animal welfare.[11]

In other words, the European Union is legally constructing limits for what Arctic seal hunts *should* be like in order to fall under the indigenous exemption of the EU Seal Regime. Special attention must be given to the notion of 'commercial seal hunts' in this context. The indigenous exemption does not allow seal hunts by indigenous peoples to be driven by commercial intent to yield products eligible for the European market. Although the exemption does refer to 'income to support life and sustainable livelihoods',[12] thus indicating the generation of money from the sale of seal products, the distinction between seal hunts for subsistence and commercial purposes is not easy to uphold.[13] Moreover, not perceiving the Inuit as being embedded in modern economic systems neglects substantial *in situ* realities in the Inuit regions of the USA, Canada, Greenland and Russia.

As a result, Inuit and other groups have launched several court cases before the European Court of Justice (ECJ) in order to overturn the ban. Although legally exempt from any trade barrier, the interlinked trade pathways of Inuit and non-Inuit seal products, the overall abstention of buyers from purchasing any seal products, and the market economic realities in Inuit communities led to drastic impacts on Inuit communities. Although being ultimately unsuccessful, the court cases demonstrate the empirical limitations, and arguably imagination-shaped character, of legal rules regarding Arctic economies and livelihoods.[14]

Arctic livelihoods as a legal construct

While the Arctic finds many reflections in international legal regimes, the focus of this chapter is 'Arcticness', a concept which this author understands to contribute to a better comprehension of what Arctic realities encompass. Many attempts have been made to decipher the scopes and limits of 'the North', the 'Circumpolar North' or the 'Arctic' and up to this point no fully satisfactory definition, at least for this author, has been found.[15] Contrary to what the Arctic as a geographical region entails, despite the ambiguities surrounding the definition of 'Arctic' and, in particular, 'culture',[16] Arctic cultures appear to be more clearly defined legally by law-makers. Conditions are inserted into legal regimes that lay down conditions for what constitutes a 'real' Arctic culture and what does not. Along with these criteria go the conditions for the delimitations of Arctic economies which, as shown by Glømsrod and Aslaksen, are diverse and complex in nature.[17] First and foremost, these legal reflections of Arctic narratives unveil significant shortcomings in knowledge regarding livelihoods and economies of contemporary Arctic societies. At the same time, legal regimes now attempt to include Arctic and other indigenous cultures, their traditional ecological knowledge (TEK) and livelihoods as a valuable element in sustainable development and environmental protection.

For example, the Convention on Biological Diversity (CBD)[18] in its Article 8(j) lays out that the contracting parties are to 'respect, preserve and maintain knowledge, innovations and practices of indigenous and local communities embodying traditional lifestyles relevant for the conservation and sustainable use of biological diversity'. While this is held very broadly and is not limited to Arctic indigenous peoples, it becomes more concrete in a Canadian context where, for example, the Oceans Act[19] in Article 42(j) stipulates that traditional ecological knowledge is to be used 'for the purpose of understanding oceans and their living resources and ecosystems'. Procter questions how far this approach towards TEK reflects the interests of the knowledge holders and argues that the discourse surrounding TEK and its embodiment in law ultimately reflect neo-colonial approaches given the inherently exploitative nature with which TEK is utilised.[20] Thus, conceptualising the cultural practices and knowledge systems under the banner 'traditional ecological knowledge',[21] while well-meaning, neglects the cultural diversity of indigenous peoples, simplifies the socio-cultural attachments to the knowledge as well as negating the achievement and generation of this knowledge.

A deeper analysis of 'knowledge' in the Arctic and elsewhere, particularly in environmental governance, furthermore reveals a systemic bias in approach: although the CBD explicitly refers to 'indigenous and local knowledge', by and large the terms 'traditional' and 'local' are equated with *indigenous* knowledge holders. The same, I argue, is prevalent in the notion of 'tradition' or 'traditionality'. Indeed, Berkes exemplifies this with reference to the West Indies and notes: 'Strictly speaking, the West Indies is one part of the world in which traditional systems do *not* exist' since 'the indigenous populations … have almost completely disappeared'.[22] This approach, as Berkes himself acknowledges, neglects that vast ranges of long-standing systems that have developed (or are developing) based on which management systems have evolved or are evolving. Inevitably, whether or not those engaged in the development of these systems are considered indigenous or not is epistemically irrelevant.[23] Approaching Arctic socio-cultural systems by making a distinction between 'indigenous (knowledge) systems' and 'non-indigenous (knowledge) systems' creates a bias which points towards a 'museified' perception of Arctic living conditions.[24] In other words, this approach neglects, first, the presence and knowledge of non-indigenous Arctic residents; second, the close intermingling of indigenous and non-indigenous people and peoples in the Arctic;[25] and third, the 'modernisation' of Arctic indigenous economies, which concerns the shifting towards a market economy.

'Traditionality' of Arctic livelihoods therefore refers to livelihoods that are perceived as pre-colonial. The 'indigenous exemptions' utilised in regulating the marine mammal hunt or the trade therein reflects this stance. Two points must be made in this context. From a legal perspective, this 'museified' view came to play a role in the *Länsman v Finland* case that was brought before the Human Rights Committee (HRC). The HRC oversees the implementation of the two core human rights covenants.[26] The HRC argued that:

> The right to enjoy one's culture cannot be determined *in abstracto* but has to be placed in context. In this connection, the Committee observes that article 27 does not only protect *traditional* means of livelihood of national minorities .... Therefore, that the authors may have adapted their methods of reindeer herding over the years and practice it with the help of modern technology does not prevent them from invoking article 27 of the Covenant.[27]

This author would argue that preventing indigenous peoples' access to a resource and access to the market for that resource due to the modernisation of practices and technology, stands in violation of the finding of the HRC. The HRC comment is, indeed, a landmark comment on acceptance of the technological development of indigenous peoples within the context of 'tradition'. The legal response, however, for example in the European Union, has been slow. In all fairness, the EU does not consider subsistence hunting as being merely limited to the community sphere of exchange[28] but as holding an external dimension. Otherwise, the notion of 'income support' would not appear comprehensible, nor would an 'indigenous exemption' be necessary in the first place. Notwithstanding, the law banning the trade in seal products in the EU narrates Inuit culture and economy in an outdated fashion. Although the Inuit were consulted as part of the preparatory process of the EU Seal Regime from 2006, by and large the perception of Inuit seal hunts still reflects the narratives of 'traditionality' as in the regimes of more than 100 years ago: the hunting of seals with 'traditional' hunting gear such as harpoons; the hunting of seals from canoes and/or on the ice; the hunting of only very small numbers of seals; and the processing and utilisation of seals on site without a larger external dimension.[29]

In Greenland, albeit a part of the kingdom of Denmark but having left the European Community formally in 1985 while Denmark remained a member, the Arctic and its livelihoods play an integral part in the legal environment. This is particularly the case with regard to the hunting of marine mammals. Since its colonisation, Greenland's legal system has been subject to a dual system of Greenlandic and Danish rules – for example, the Greenland Administration of Justice Act in 1951 – but the legal and court system on the island has become significantly more 'Danisized'.[30] Hunting for any species in Greenland has therefore been subject to stringent national and international legislation. Since Greenland's population is considered indigenous by the international community, any hunting for marine mammals falls under 'indigenous exemptions', both by the EU Seal Regime as well as under the ICRW. The hunts conducted, therefore, are considered legitimate and justifiable as fulfilling subsistence – and not commercial – needs. While this is normatively the case, the legal regimes in Greenland strictly regulate marine mammal hunts. By taking a microscopic look at the provisions of the legislative framework, little consideration for the realities on the ground, or for Arctic livelihoods, appears to be embedded therein. Licensing, animal welfare, reporting, technological requirements and quota provisions, to name but a few aspects, are an inherent part of the legal environment in Greenland. In terms of content, not structure, the Greenlandic system thus resembles other states in which marine mammal hunts are conducted, for instance Norway.[31] Only by taking a step back can the consideration of Arctic livelihoods become apparent. This is particularly the case in international terms, for instance in the view of the IWC. Greenland's representatives vehemently defend its marine mammal hunts and frequently refer to the necessity of whale and seal hunting for the benefit of Greenland's people.[32] Moreover, Greenland's membership in the North Atlantic Marine Mammal Commission (NAMMCO), a 'sustainable use' organisation, indicates its position as defending its right to hunt marine mammals.

From a legal perspective, this puts Greenland in a difficult position. How is it possible for the island to defend its 'Arcticness' as part of the Kingdom of Denmark, which, in turn, is part of the EU that holds a stringent anti-sealing and anti-whaling stance? Crucial in this regard is Declaration 25 to the Maastricht Treaty of 1992, which enables Denmark to diverge from the EU's common position in the interest of Greenland and the Faroe Islands.[33] Denmark first invoked the Declaration at the IWC meeting in 2008 and intervened on behalf of Greenland and the Faroe Islands and contrary to the EU's common position.[34] In addition, at the 2016 meeting of the IWC, which the author attended, Denmark defended the interests of Greenland in particular. For example, although no quota was allocated to Greenland's Aboriginal Subsistence Whaling hunts for 2013/14, several whales were taken by Greenlanders. Several anti-whaling nations considered this an infraction and stated that the act should therefore be treated as such. Denmark, however, disagreed with this assessment and explicitly referred to the subsistence needs of Greenlanders that were met by taking a small number of whales.[35] Whether or not the EU took the same view cannot be ascertained, but given the EU states' normative alignment with the anti-whaling nations in the IWC, it can be presumed that the EU would consider the hunts an infraction. Indeed, Denmark surfaced prominently at the IWC meeting and openly represented Greenlandic interests. Although the author did not attend the coordination meetings of the EU, Greenland's arguments appeared to imply they considered Declaration 25 to still be valid. Or to put it differently, Denmark openly voiced its support of Greenlandic whaling given its inherent part of Greenlandic culture and livelihood.

'Arcticness' in other legal contexts

It goes without saying that the Arctic and 'Arcticness' play a significant role in other contexts, particularly as regards climate change mitigation and adaptation. These issues will only be touched upon here very briefly. In general, the significance of the impact of climate change on the Arctic region shifted onto the world's agenda after publication of the seminal *Arctic Climate Impact Assessment* (ACIA)[36] in 2004/05. After release of the report, the Arctic Eight all released their respective Arctic strategies in which they committed to different goals while at the same time tackling climate change by reducing greenhouse gas emissions and fostering 'green' economies. Moreover, the ACIA report helped to bring the Arctic more closely into the scope of the international climate change regime. Since the third and fourth assessment reports of the Intergovernmental Panel on Climate Change (IPCC) in 2001 and 2007 respectively, the Arctic has become an integral part of the regime.[37]

Although, from an environmental perspective, this appears to correspond with the interests of Arctic peoples, the situation is slightly more complex. The role of Greenland in the climate change regime is twofold: on the one hand, Greenland's population depends on the Arctic environment and suffers greatly from melting ice.[38] On the American side, a legal expression of 'Arcticness' was uttered when in 2005 the Inuit Circumpolar Conference (now Council) filed a petition to the Inter-American Commission on Human Rights over the emissions of the United States, constituting human rights violations due to climate change implications. Similarly, the Arctic Athabascan Council filed a similar petition over Canada's black carbon emissions in 2013. Although Arctic indigenous groups, in particular, frequently highlight the adverse effects of climate change on the natural environment, Sejersen presents a stunning insight into the 'other' side of the climate change debate, namely the new opportunities, for example as regards hydrocarbon exploitation, for Arctic communities. This is especially the case in the context of Greenland's aspirations for independence. Sejersen shows how industrial mega projects have now become a political and economic reality in Greenland, which is often still portrayed as a remote and somewhat backward orientated island in the Arctic.[39]

While the above discussion appears to imply that the legal reflection of 'Arcticness' is limited to environmental factors such as resource abundance or climatic changes, a brief look should be taken at soft-law means of cooperation in the north.[40] The Barents Euro-Arctic Council stands out in this regard as this north-eastern European organisation goes much further than environmental cooperation. 'Arcticness' or a common identity as part of the Barents Region has spawned cooperative structures for education, health and social issues, transportation, youth or investment.[41] While this cooperation is not based on legally-binding documents, but on declarations and Memoranda of Understanding, the Council reflects the regional understanding of the legal 'Arcticness', inevitably dealing with more tangible issues that have direct and immediate effect on the people in the region.

Conclusion

In the context of marine mammal hunting, in particular, the concept of 'Arcticness' is primarily linked with specific understandings of Arctic peoples and economies. These understandings largely correspond to those narratives that were applied in the early regimes more than 100 years ago that regulate and manage marine mammal hunting activities. This view on the Arctic therefore romanticises and 'museifies' the Arctic with little regard for the socio-economic conditions that have arisen over the last 50 years or so. The legal frameworks reflect these narratives and create a legal space for Arctic cultures with significant impact on living conditions in the high north. As a result, Arctic peoples appear to be put in boxes from which it is difficult for them to escape. Only through targeted 'securitising moves'[42] that use the legal and political environment in which Arctic peoples are located can these narratives be challenged. These moves occur through action taken by Arctic people themselves, as in the examples of court cases before the ECJ or Inuit and Athabascan petitions, or through their official representation in international fora, such as the IWC via, for example, the Danish representation.

From a legal perspective, it can be concluded, 'Arcticness' lies in the eye of the beholder. This chapter has taken a somewhat critical approach towards the outside view – imagination – of the Arctic and its reflection in Arctic-relevant laws, such as for marine mammal hunting. And it was argued that this imaginary perception of 'Arcticness' that has found its way into legal regimes actually causes hardships for Arctic communities by not taking real-life circumstances adequately into consideration. More research is needed on the anthropology of Arctic legal regimes to investigate the origin of Arctic legal understanding and the impact on contemporary Arctic societies.

PART 3
Arcticness Futures

11
Continental divide: Shifting Canadian and Russian Arcticness

Nadia French, Mieke Coppes, Greg Sharp and Dwayne Menezes

Introduction

People are often defined by the locations in which they are born, in which they live or to which they are culturally and historically tied. Geography is one of the many factors that plays a role in shaping culture, politics and society. Yet, societies are tied to more than one defined geography which can have implications on their development. The Arctic is but one example. Generally perceived as one region – and in many cases, with having one identity – the Arctic holds eight countries and several indigenous nations with distinct cultures. Sometimes, Arctic provinces seem to have more in common with each other than with other areas of their countries; however, it is misleading to assume that these commonalities mean that the Arctic is changing uniformly. It should be clarified that the authors believe that there are no absolute differences or similarities between geographical regions considered in their socio-political representation and this chapter only seeks to outline and propose the instrumental benefits of such a theorisation.

The future of the Arctic is fundamentally important to the future of the world, not only due to the fact that a changing climate means a changing world, but also because the north teaches the world about resilience and the importance of working together. The peoples who have been living in the Arctic know that it is not only beneficial to work together and protect the environment, but it is part of the very core of life. Without this, there would be no ability to live successfully in the Arctic.

Speaking of the future of the Arctic can conjure many different aspects, including the changing climate, the impacts of tourism, potential new shipping routes, extractive natural resource development and developing infrastructure needs. Each of these aspects plays out on the international stage in a plethora of ways, but the string that ties these pieces together is the people who live there, those who are being directly impacted by the issues. Although the region is changing dramatically, in some ways universally, the reaction to and the implications of these changes are different. Across the entire region, the climate is changing and ice is melting, which means the opening of shipping routes and a potential increase in both long-term and short-term populations. Nevertheless, how countries react to this, how their politics and society are shaped by these changes, is different and dramatically so in some cases.

This chapter will analyse the two countries with the largest geographical space in the region and assess the shifting political and societal changes in both Canada and Russia, providing a window into how two countries are reinterpreting their relations with the north. Canada and Russia have historically-rooted differences that have led to the creation of two Arctics: culturally, socially and politically. Furthermore, the core political and social values of these countries will be assessed through the lens of their Arcticness, generally defined as a perceived right of a state to the Arctic territory, while also evaluating north–south relations within the countries. It is clear that 'discussion on the changing Arctic environment, as well as on the impacts of such change on the cultures and livelihoods of indigenous and local communities, plays out against the backdrop of the shifting views on the concept of sovereignty in international relations and international law'.[1] The role that sovereignty and the view of self plays in how countries and Arctic territories develop should not be negated. This is why, throughout each section of this chapter, three areas will be expanded on to highlight the present and future trajectories of Arctic development in the context of a narrowing international gaze on this part of the world. The political and social dimension of north–south relations will be analysed from the perspective of historically defined relations between the colonisers and the colonised. Particular attention will be paid to the different meanings these relations have produced, which appear idiosyncratic to each continent.

Canadian Arctic

Canada's north has played an important role in the creation of Canadian national identity. The region is still vaguely defined in three main ways: as north of the Arctic Circle, north of the 60th parallel, or north of the 60th parallel and the Inuit homeland, Inuit Nunangat. For the purpose of this analysis, the definition of Canada's Arctic will be the political definition of the term and, therefore, it will include not only the geographical region above the Arctic Circle, but also the entirety of the three northern territories and the Inuit territories in Québec and Newfoundland/Labrador (Nunavik and Nunatsiavut respectively). This distinction is important to note, due to the fact that although parts of Canada's north are not geographically in the Arctic, the culture and society found in these regions are intrinsically tied to the Arctic and, therefore, are integral in any analysis seeking the implications of change in the region.

The political rhetoric and culture in Canada's Arctic has undergone dramatic shifts over the last 50 years. This, in turn, has had implications for both those who have lived in the region for millennia as well as newer transplants. This changing society is still deeply rooted in the historical bearings of the region, which creates a dichotomy of old and new, of indigenous and settler, and of sovereignty and multinationalism. Some of the Canadian north's most fundamental changes in the last few years have been political, demographic and a shift in the state of knowledge and its intergenerational transition.

Canada's Arctic has always been felt as an important region in the psyche of the Canadian people. From the role of the polar bear to the northern lights, from the use of the Inukshuk to the international role that Canada plays as a 'Northern' nation, the Arctic is inherently important to the people who live there and important to how Canada sees itself on the international scale. This claim to 'Northernness' could be construed as disingenuous, as the vast majority of Canadians do not live in the Arctic, and most have never visited. In fact, of the approximately 33 million people who call Canada home, only approximately 104,000 of them live above the 60th parallel.[2] Even with so few people living in these northern regions, often in very challenging conditions that differ from much of the rest of Canada and with different issues than the rest of Canada, the dramatic changes that are happening there cannot simply be shrugged off as regionally 'Arctic' issues, but must be recognised in their intrinsic Canadian and Arctic nature. The three major shifts – political, demographic and knowledge-based – that will be analysed in this chapter tell a story of a region that is simultaneously Canadian and Arctic, yet neither at the same time. It is a region that can be defined neither by the country of which it is a part, nor the geographic space to which it belongs.

Political change

Changes in the discourse of Canadian politics are one of the key issues surrounding the north. Previous Prime Minister Stephen Harper created a specific political rhetoric when it came to the region, which focused on sovereignty and militarisation. 'From attempting to replace the beaver with the polar bear to substituting human rights leaders with icebreakers on the fifty-dollar bill, Harper has used his time in office to determinedly shift Canada's national identity from a "peacekeeping nation" to one focused on security and strength', with the north playing a crucial role.[3]

With the election of Prime Minister Justin Trudeau, there was an expected change to a more Canadian view on collaborative multinationalism. The new politics 'mark a return to Canada's historic emphasis on multilateralism and careful diplomacy. Indicative of as much, the relationship with Russia has stabilised after Trudeau's government took over in November of 2015'.[4] However, the Prime Minister has yet to make known his specific politics in the Arctic, and there is much uncertainty as to the long-term implications of a Trudeau government. The difficult relationship between the indigenous peoples in the north and the federal government, located far to the south, has led many to question the policies implemented by Ottawa. This is unsurprising given the colonial history marred by discrimination, institutionalised abuse and forced assimilation. Experiences such as these were not uncommon: 'Nomadic hunters were forced off the land into settlements. Children were sent to residential schools in the south. Tongue-twisting names in native languages were discarded in favour of numbers. Social problems, such as rampant alcoholism and drug use and a high suicide rate, were rife in the settlements. When Canada felt the need to assert its sovereignty in the 1950s, Inuit families from northern Québec were relocated to unfamiliar terrain in the high Arctic. Many of these "human flagpoles" grew sick and died'.[5]

This history informs the politics of today and the mistrust that is often still prevalent. It also ensures that many stay sceptical about the political promises that are being made, leading people to ask: will policy shift as well as rhetoric? Prime Minister Trudeau's rhetoric is focusing on nation-to-nation relations with the indigenous peoples in Canada, which can in part be seen by the appointment of Mary Simon as the Minister of Indigenous Affairs' Special Representative on the Arctic and the official adoption of the United Nations Declaration on the Rights of Indigenous Peoples (UNDRIP).[6] However, people are still waiting for the real policy applications of this changing rhetoric: UNDRIP has yet to be incorporated into Canadian law and discussions are ongoing. Canada is currently at a crossroads in its political relationship with its northern half. Only time will tell how the real day-to-day international and domestic policy will shift in the coming years.

Changing demography

Canada's Arctic has always been vast, but remote from the south. Although the percentage of indigenous peoples is particularly high in the north, especially in the Inuit territories of Nunavut, Nunavik, Nunatsiavut and Inuvialuit, this is not true for the whole region. Yukon, for instance, is predominantly non-indigenous with only about 20 per cent of the population being First Nations.[7] This highlights the problems of making generalisations even within Canada's north, let alone across the entire Arctic region.

There is now a shift in the demography of the north, not only in the nationalities of people living there, but also in the average age of the people living there. The latter change will be fleshed out in the following section. According to the 2011 Canadian Census, there are approximately 2,900 Canadian immigrants living in the Northwest Territories, with a large swathe coming from the Philippines, the UK, Vietnam and the United States.[8] The population of the territory, which is an area of approximately 1 million km^2, is approximately 43,500 people. That means that approximately 6.7 per cent of people living in the territory are immigrants. According to the 2011 National Household Survey, 11.3 per cent of the population of Yukon were foreign-born or immigrants, with the largest percentages coming from similar countries as the Northwest Territories.[9] Nunavut was the territory with the lowest proportion of immigrants at only approximately 2 per cent of the population.[10]

These numbers show a changing landscape in the north. With more international immigrants, and some refugees, heading to the north, some of the societies in these regions are also in flux. This is part of a larger trend in Canada which is attempting to successfully create a multi-national, multi-cultural society. Although the numbers of immigrants are not extremely large in the north at the moment, with the changing climate and the potential for milder weather, the Canadian north may have to prepare for a large shift in the demographic which will have lasting impacts on its societies. The Northwest Territories, for example, recognises the benefit that immigrants can play in boosting the economy and building on the society in the region. The Nominee Program, which is one example of a government policy attempting to achieve this boost, is designed to ensure high-skilled individuals, including immigrants, are living and working in the Northwest Territories.[11] As Jackson Lafferty, Deputy Premier and Minister of Education, Culture and Employment, said in a speech, '[The Nominee Program] initiative is a key component of our Growth Strategy, aimed at attracting 2,000 new residents to the NWT over the next 5 years'.[12]

A large population of indigenous peoples still live in the north. From the four Inuit territories, to the Dene people in the Northwest Territories, to the Gwich'in in the Yukon, there are many First Nations, Inuit and Métis peoples living above the 60th parallel. This does not mean that the region is not diverse: according to the Northwest Territories Language Commission, for example, the Government of the Northwest Territories recognises 11 languages as official, including English, French, Cree, Inuktitut and Gwich'in.[13] As shown by the differences in language, the differences among indigenous peoples in the north should not be negated. Indigenous peoples in Canada include a vast array of First Nations, Métis and Inuit, each with their own history, culture and traditions. Although there will likely be an increase in the immigrants moving to the northern region of Canada, the society there is already varied in many ways.

The changing state of knowledge

The way knowledge is passed down from generation to generation was almost entirely disrupted in the twentieth century due to the horrors of residential school. The experiences in the north were somewhat different to the rest of Canada partly because of the remoteness and the lack of economic development in the region at the time. In fact, there were only six residential schools in the three northern territories by 1950. This demonstrates that speaking of such an issue on a country-wide basis can be misleading, so much so that the Truth and Reconciliation Council (TRC) wrote a separate report on 'The Inuit and Northern Experience'.[14] This Commission was designed as a response to the residential schooling and the healing that was needed. The report that came out of the TRC was not only an analysis of the legacy of residential schooling, but also an indication of how Canada could work toward a healthier and stronger future together.

There were significant differences between the residential schools of the south and those of the north, notwithstanding the distances involved, but also the fact that the schools were administered (after the 1970s) by the northern governments themselves, as opposed to Indian Affairs, which was the case for the southern schools.[15] These schools, even though they were not segregated, 'disrupted the intergenerational transmission of values and skills and imparted few if any of the skills needed for employment'.[16] Not only was there a dramatic impact on the skills transmitted, but 'when [the students] returned to their communities, they were estranged from their parents, their language, and their culture'.[17] This left a gap in communities that passed information in a way that was different to the 'Western' system; for the northern communities it was one based on the importance of the land and learning from previous generations. The loss of culture and language, and the estrangement from families, had an impact on education and subsequently the lifestyles of those living in the north.

The shape of education in the north is changing, although the underlying principles are not: 'schools are relatively new to many indigenous communities, but community responsibility for the education of the young is not'.[18] And the impacts of a Western-based education system has left questions and problems surrounding the way that children are being taught. Zebedee Nungak, who was President of the Makivik Corporation in the 1990s as well as an important negotiator in the James Bay and North Québec Agreement (a land claims agreement signed in 1975 spanning much of northern Québec), speaks to this, comparing the current Nunavik education system to the failure of the Franklin expedition. He also emphasises the societal turmoil that came from such a dramatic shift in education from one generation to the next, asking the reader to 'consider that our grandparents, the first generation of Inuit to observe their grandchildren (us) being herded into uni-lingually English federal schools, were the last of countless previous generations to leave the nomadic lifestyle'.[19] Education in the north was, in many cases, information passed down through generations, not something that was taught in a school: 'Inuit education did not traditionally comprise a separate set of practices, supervised and documented by an administrative body, this topic necessitated input from Elders who were raised and educated by their parents on the land'.[20]

With a shifting age demographic, as well as the boom of technology, the way that knowledge is being transmitted has also been impacted – not only the passing of knowledge, but also the culture and the society that depend on these relationships. Technology has a large role to play in how new generations are learning, and although the technology in the north may be slower than some southerners are used to, it still has an important role in the lives of newer generations. In today's society, the young can turn to Google instead of their elders and parents to learn the answers to some of the questions they have, and this will likely have a lasting impact on historical relationships built on learning and sharing of knowledge. The exact nature and magnitude of the impact this will have on communities in the north has yet to be determined, but one can be sure that it will shape the society of the next generations of Northerners, much as it will those living in the rest of Canada.

Russian Arctic

The place of the Arctic, and the north in general, in Russian history and national identity is punctuated with periods of heightened political interest and exploratory ventures. The populating of the Russian north began in earnest in the nineteenth century, while industrial development began in the 1930s–1940s and continued with the discovery of oil and gas in the 1960s–1980s. Seen from the south as a northern frontier, a resource bed of hydrocarbons and marine resources, a curse for convicts and a source of pride in the popular imaginary, the Russian Arctic defies a single definition. The Russian Arctic is also referred to by southerners in more abstract terms as 'a condition' of Northernness,[21] a geocultural non-place expressed linguistically through the concept of *Russkiy Sever* (the Russian North),[22] or a socio-cultural entity defined as a vernacular mental cultural region.[23] The ambiguous attitude of the Russians to their northern region is captured to an extent in the poll results of the *Fond Obschestvennoye Mnenie* (2015) which found that two-thirds of Russians support the state's policy of exploration in the Arctic; yet, the majority of the respondents to the poll expressed no desire to go there themselves.[24]

The politically-defined Russian Arctic, known as the 'Russian Arctic Zone', is generally described as a macroregion (which is defined in Federal Law on State Strategic Planning of 28 June 2014)[25] that is, a special area of state governance implying similarity of economic and political interests (and naturally that of geographical conditions). Little affinity and lateral economic or political interactions between Arctic territorial units have been identified,[26] while most of the resources required for regional development are being outsourced (from workforce to technologies to energy to food).[27]

The 'Russian Arctic' is often used, especially in International Relations studies, interchangeably with 'Russia' or 'Moscow' to indicate the Russian government.[28] This implies subjugation and inseparability of the region from its metropole. While the term Russian Arctic Zone, used extensively in domestic political documents, is itself indicative of such centre–periphery relations and the supremacy of the government in representing and managing the Russian Arctic as a border region, there are indisputably more layers to the Russian Arctic (or any political region) that are often ignored 'to accommodate the story'.[29]

Political change

After the Arctic Strategy (Foundations in 2008 and Strategy in 2013)[30] was issued, the heuristic parallels with the Soviet Union's industrialisation efforts were inevitable.[31] Russian academics (e.g. A. Granberg, A. Tatarkin, A. Chilingarov) *en masse* supported Putin's undertakings, and some even insisted that there cannot be continuity between the Soviet Union and Russia, for the former's aggressive approach and reliance on convict labour is incompatible with the apparently liberal and democratic conditions of present-day Russia.[32] Similarly, the institutional model of hypercentralisation adopted by the Soviet state and enacted through Glavsevmorput (semi-militarised 'fiefdom') and Dalstroi (industrial complex heavily reliant on convict labour)[33] have been 'succeeded' in 2015 by a mere coordinating committee with no budget of its own, located in Moscow and headed by Dmitry Rogozin.

Cartography has long been interpreted as an associate of power and domination.[34] In this respect, the new Russian Arctic map presents an interesting case study. The geographic delimitation of the so-called Russian Arctic Zone (Presidential Decree of 2014) was not based on ethnic distribution borders, nor administrative borders of subregions, nor even on the Arctic Circle of latitude, leading to the reduction of the overall area compared to the previous delimitation document of 1989. Geographic determinism of economic and political priorities of the state hit the predominantly indigenous Sakha Republic especially hard as it saw eight of its districts (*ulus*) dropped from the list of the Arctic land territories and denied investment privileges as a result.

Development and demography

Whether continuous or divergent, Russian Arctic development brings to light similarities in the challenges faced by the Kremlin before and after the regime change and the north–south dynamics created to solve them. It seems likely that in Russia as it is today, neither development nor governance of the Russian Arctic region can be fully self-sourced, that is, based on its indigenous populations and local resources. In terms of administration, in Russia, most of the Arctic provinces are governed by either first generation locals of non-native descent or southern-born and educated migrants (Chukotka Autonomous Okrug, Yamal-Nenets Autonomous Okrug, Norilsk, Arkhangelsk); only one self-proclaimed ethnic native (evenk) is heading a Sakha Arctic district.[35] Since most of the indigenous peoples have adopted traditional Russian names and speak the Russian language, it becomes nearly impossible to determine the ethnic association of a person living in the Arctic without overt self-identification or direct inquiry. It is further complicated by the fact that only about 25 per cent of the total Russian Arctic population is made up of ethnic natives, with Russians representing the unrivalled majority.

Yet it is often overlooked that 'colonisation' of the Arctic is not a one-way street: the non-indigenous settlers and their descendants have in the past undergone so-called 'indigenisation' (in Russian, *korenizatsiia*),[36] whether through intermarriages or by self-identification. At the same time, some natives abandoned traditional lifestyles or migrated to cities, other parts of Russia or abroad. Marina Kovtun, the Murmansk-born governor, noted that being a Murmansk citizen is a 'trait of character, of the soul',[37] pointing at multi-culturalism, on the one hand, and the values of national unity, on the other.

The workforce required to effectuate any industrial project and the fluxes of migration from such industrialisation have already incited several discussions on the complex socio-economic development of the region, the future of Siberian monotowns and workforce supply strategies.[38] The total deficit of skilled workers in the Russian Arctic, according to official sources, amounts to 25,000 people a year.[39] This number obviously does not cover the available migrant workforce or factor in infrastructure and auxiliary personnel as well as unemployed family members that breadwinners bring with them. The general migration trend, however, remains negative with the Russian north and east losing population to the western and southern regions with some exceptions (e.g. Yamal-Nenets Autonomous Okrug).

Shift work and the expedition method of exploration in the Arctic have gained ground as strategies for cutting infrastructure costs in the most recent remote north Siberian projects (e.g. 'Yamal LNG', liquefied natural gas). They have made it more difficult to assess the size of the external element of the population and to evaluate the extent and type of social impact this has on indigenous communities and the natural environment.

The changing state of knowledge

The northern ethnic groups are divided into large indigenous groups (i.e., over 50,000 people) and small-numbered peoples of the north, of which there are 17 in the Russian Arctic as per the Law of 2015 on the Small-Numbered Indigenous Peoples,[40] although only the latter have a special protected status. Assimilation of the indigenous population was a result of extensive economic migration to the north in the twentieth century and aggressive interference by the Soviet government in the economic, political and cultural practices (e.g. Resolution of the RSFSR Ministers' Council of 1960 on 'Additional assistance in economic and cultural development of the peoples of the North').[41] Nonetheless, the Soviet experience was not all negative: ABC-books in local languages were first published in the 1930s; the teachers that taught in indigenous communities were recruited from the indigenous peoples; and nomadic schools, too, first appeared in the 1920s.[42] Today, the nomadic form of education is being tested in Yamal and Yakutia, but special boarding schools still remain the most widespread form of primary and secondary education in the remote parts of the north.

The approach to ethnic policy-making in present-day Russia changed, but some of the problems (e.g. maintaining the balance between traditional culture preservation and culture-sensitive modernisation) still remain. In 2016, the Government of Russia signed a plan of action for the third and final stage of the Concept of Sustainable Development of Small-Numbered Peoples of the North, Siberia and the Far East (2009) for 2016–2025.[43] The Concept of Sustainable Development (2009) foresees a list of loosely-defined measures intended to improve the standards of living and update regulations related to state support of indigenous peoples (e.g. fishing and hunting rights, internet, ethnic tourism, transport services, power supply, employment stimulation, alcohol restriction and distance learning).[44]

Quintessentially, the contemporary ethos of indigenous population development can be found in the words of the Yamalo-Nenets Autonomous Okrug Education Department Director on nomadic schools: 'We must give parents and children a right of choice' to stay in the community and live a traditional nomadic life or seek a modern life.[45] This choice depends not only on the schooling system, but also on the state of the natural environment, limitations of industrial development, and inclusion in the decision-making process and profit-sharing from the industrial use of land. The plan under the 2009 Concept is to be realised before 2025; however, a large amount of industrial construction is already under way.

Siberian (including Arctic) autonomous okrugs have been granted to the indigenous peoples since the 1930s to accommodate their right to self-determination; the irony is that the okrugs located within the oil, gas and other natural resource regions have attracted external economic actors and political leadership that co-opted cultural identification for the purposes of gaining political weight.[46]

Differences and similarities between the Canadian and Russian Arctic regions

Canada's and Russia's northern frontiers both experienced a period of southern discovery, geographic exploration, colonisation, resource boom and migration. Now, both face many of the same environmental crises and some of the same political questions that arise as a result. The ideology that accompanied these endeavours often differed, yet both countries had dialogues with indigenous cultures and were transformed by them. Similarly, the mentality towards and the relationship with the environment came full circle, starting from a perceived emptiness to the current recognition of the complexity that is essential for survival of not only the indigenous peoples but also for the entire human race.

Both Russia and Canada seem to have a vague definition of what the 'North' truly means to their country, geographically, politically and culturally. This stems not only from an historical distancing of the region and a preference for southern views, but from a more current recognition of the overarching impact that the 'North' can have in defining the country's self. Both Canada and Russia are internationally recognised as being Arctic countries, often with the perception of them being countries of the ice and snow. This recognition and the assumptions of what the Arctic is in the perceptions of others shapes the role of self that these countries have created.

Russia's multi-ethnicity never experienced colonialism *sui generis* and, therefore, never had an emergence of the post-colonial discourse that has been seen in Canada.[47] Although Canada's north is also multi-ethnic, the settler identity struggle is still seen in modern day discussions, both politically and culturally. Canada continues to strive for a 'North' that manages to overcome the dichotomies that have presented themselves openly and often. The shifting political rhetoric in Canada, as well as the further inclusion of indigenous peoples on the territorial and also national scale should not be ignored. Russia, on the other hand, does not have open political confrontations between the centre and the north. But this does not mean that there is no problem of cultural domination – the fact that post-Soviet democratisation was insufficient for indigenous peoples to fully reclaim their traditional names is a telling example of the pressures that Russian society inflicts on its mostly small-numbered indigenous communities. Similar to the Soviet period, in contemporary Putin's Russia, clandestine 'grey' politics – such as threats, bribery of tribe leaders, ambiguous laws, whistleblowing, nepotism and more – is still practised in every sector where there is a conflict of interests, including indigenous rights to territories and natural resources, at least according to the indigenous leaders themselves.[48]

What is especially striking is that both countries seek national unity, through a strengthening of the vertical relations in Russia and the new federalism in Canada. For the latter, it means empowering the Arctic peoples, while fostering an openness for dialogue and multinationalism. For the former, it means tightening ties between the south and the Arctic through migration, development and government.

Different structures of the population between Canada's and Russia's Arctic raise issues of differentiated regional governance and direct versus remote influences of the south on the day-to-day life, identity and inter-ethnic relations. Arctic units in Russia are not monoethnic in the sense that several Arctic peoples can share the same region with sub-Arctic or non-Arctic settlers. Canada's Arctic is difficult to define in a similar way due to the vast differences in the regions, both physically and culturally. While some regions may be predominantly Inuit or First Nations, many of the regions in the north have a wide range of Inuit, Métis, First Nations, non-indigenous, immigrant and other Canadian within them. This brings a host of problems and opportunities within it vis-à-vis the governance structures of the regions, for which solutions are being sought at the local, territorial and national level. For example,

Deline, a community in the Northwest Territories, has recently created the first combined indigenous/non-indigenous government in the territory, which is designed to ensure that all people, those who are and are not Deline First Nations, are equally represented.[49]

There are similarities in historical approaches to education as a means of accelerated modernisation, which are not unique to the Arctic region, but the old habits in this area prove to be surprisingly tenacious. This can have lasting implications for the cultures that depend heavily on the transmission of knowledge from one generation to another as a means of developing bonds and constituting an integral part of the society. Furthermore, modern technologies, such as mobile phones, computers and snowmobiles, now widely used by the indigenous communities across the two hemispheres, create a demand for new skill sets. Additionally, industrial development in the Siberian north may create other kinds of economic and cultural pressures on the local indigenous peoples.

The Arctic in the modern world often seems inseparable from its ruling state not only politically, but technologically too. For many, it may feel as though it is locked in the path-dependent trajectory of state–Arctic relations. In that sense, circumpolar fora that bring together Arctic states paradoxically recreate the same pattern. Thus, the concept of a single Arcticness, attractive as it is, is closer to the *terra nullius* (as the British saw Australia and Canada)[50] than to the multitude of 'Arctics' created and re-created through continuous south–north interactions.

Learning from the continent-bound Arcticness

The Arctic has always been described latitudinally but rarely longitudinally; yet, the places within it are often defined and shaped by their vertical, north–south connections. The complexities of relations between the immigrants and the indigenous peoples in Russia and Canada – shown through policy documents, national rhetoric and identity narratives, among other media – tell a story of alternative Arctic futures.

The Canadian and Russian northern frontiers have gone through immense and drastic socio-political changes in the last 50 years, albeit for different reasons. In both cases, these changes impacted the society and politics of the circumpolar region on local, regional and global scales. In more recent years, environmental changes, which span across the region, have led to very different social and political outcomes. Not all of the changes have been mentioned here, but books could be and are written on the ever-shifting landscape that is politics and society in the American and Eurasian North, including territorial politics, devolution and the implications of colonisation and rapid modernisation on the psyche and lifestyle of the people.

Moreover, each subregion of the Canadian and Russian north has its own set of challenges, changes, opportunities and options, which are not always scalable. Some of these changes are more national in nature, while others are more global and 'pan-Arctic' in scope. Viewing the region in its north–south dialogue can uncover hidden tensions and path-dependent trajectories that cannot be addressed and resolved through a circumpolar Arctic paradigm alone. Viewing the region from a pan-Arctic lens also allows for the uncovering of commonalities, thereby reinforcing the challenges and opportunities that these Arctic territories face, in being influenced by different geographical locations and in being distinct in their identity from the rest of their nations and the Arctic at large.

In the context of the growing importance of Arcticness, there should be an awareness that persistent issues of continental divide and north–south arrangements can become more acute and yet are dismissed as a momentary obstacle in the global effort to 'save the Arctic'. Although there are changes within the Arctic that are being felt across the entire region, this does not mean that the politics, societies and cultures within each continent are dealing with it in similar ways. It is therefore important that the Arctic itself does not become merely collateral in the new political exchange between southern-based governments of the Arctic and non-Arctic states.

12

Imagining the future: Local perceptions of Arctic extractive projects that didn't happen

Emma Wilson, Anne Merrild Hansen and Elana Wilson Rowe

Introduction

External imaginings of the future Arctic range from protected wilderness to booming oil and gas province, and proponents of different visions frequently clash in global public arenas. At the same time, external perceptions, whether pro-development or pro-conservation, frequently fail to reflect the realities of living in the Arctic, or to incorporate the views (and imaginings) of local inhabitants – those most affected by Arctic resource projects.

The Arctic region does have significant resource potential. The United States Geological Survey estimated that 25 per cent of the world's undiscovered petroleum reserves were to be found in the Arctic.[1] The Arctic also represents around 10 per cent of the global nickel, cobalt and tungsten markets, 26 per cent of diamond gem stones and up to 40 per cent of the global production of palladium.[2] Yet uncertainty about the viability of natural resource projects is ever-present. Companies may be highly visible and a project intensely debated long before it is clear whether natural resource deposits, national-level negotiations and global markets will result in actual extraction for the market. Often local communities have very little information available at this point and yet the very prospect of an industrial project can transform the way a local community imagines – and prepares for – its own future.

While the challenging work of seeking equitable, just and environmentally sound practices around natural resource projects has been much studied, we know too little about the societal consequences of anticipated but ultimately unrealised projects. In this article we explore three cases of Arctic extractive industry developments – in Russia, Norway and Greenland – where a highly anticipated extractive industry development has failed to take place. We consider the local expectations around the development and how the fact of it not taking place has affected local peoples' perceptions of their future prospects.

What is characteristically 'Arctic' about Arctic extractive industries?

Given the overall theme of the volume, we reflect here on some of the commonalities of Arctic extractive industry development. Why does it make sense to review our three Arctic case studies in conjunction with one another?

One shared factor is the extreme sensitivity of the Arctic environment and the length of time it takes for damaged ecosystems to recover. By expanding further into the Arctic, extractive industry exploration is increasingly encroaching on isolated and vulnerable territories, often on indigenous peoples' lands or in the waters where they hunt or fish. This environmental and social vulnerability has drawn extreme levels of global concern about the prospect of extractive industries expanding further into the Arctic, as indicated by the campaigns of international environmental non-governmental organisations (NGOs) and indigenous rights groups.[3] The risks include climate change, which is a dominant feature of global Arctic discourses.

Second, it may make sense to compare Arctic case studies simply because Arctic stakeholders themselves make these intra-regional comparisons. Three decades of post-Cold War 'region building' in the circumpolar north make it likely that Arctic communities look first to one another for lessons learned; likewise for companies and regional governments in their planning and policy-making. Ever stronger links are being built between Arctic (and sub-Arctic) indigenous groups, sub-state regions and communities. Links are strengthened through international academic and civil society networks; increasingly strong international legal and regulatory guidelines, some of which are Arctic-specific, such as those issued by the Arctic Council; and increased use of social media. This having been said, comparative analysis between Arctic and non-Arctic regions is also extremely valuable.[4]

A widely shared feature across the Arctic is the historical tendency towards establishing single-industry or 'one company' towns ('monotowns' in Russia) at the heart of which is a single, dominant or 'town-forming' industry.[5] Single-industry towns have often faced repeated boom and bust cycles related largely to the price of commodities on global markets, and frequently leading to extreme poverty and social dislocation.[6] This type of development push is sometimes interpreted in terms of centre–periphery economic development, associated with large states and colonial or imperial expansion, where the far flung corners of a polity provide raw materials to be processed and marketed in and for the imperial or national 'centre'.[7] The government may have identified the lands where the resources are to be extracted as being 'unproductive', despite them being highly productive from an indigenous perspective.[8] An overwhelming focus on extractive industries in the political economy and development planning is sometimes dubbed 'extractivism'.[9] In the Arctic context, this has been contrasted to the indigenous cosmologies based on sustainable resource use with which extractivist policies and projects frequently come into conflict.[10]

As Arctic communities and resources have become incorporated into global capitalist markets, the focus has mostly been on large-scale high-investment development of internationally valued Arctic resources – oil, gas, minerals, timber and fish. This kind of 'single point' economic development encourages a continuation of the 'single industry' vision of twentieth-century expansion, with bold versions of the future or efforts to 'save' the community via one grand project.[11] Policy-makers in national capitals rarely envision an economic future for Arctic communities that is as complex and multifaceted as those anticipated for more southern towns and cities.

Moreover, the high cost of such ambitious, monolithic development planning in the Arctic means that a drop in commodity prices might translate rapidly into the withdrawal of investment from expensive and risk-laden Arctic environments. The oil price collapse of 2014/2015 triggered the withdrawal of a number of oil majors from Greenland, while the rise of the shale gas industry in the United States drove down gas prices and contributed to the decline in investor interest in Russia's Shtokman project (see case studies in this chapter). In rare cases, a community has the opportunity to decide themselves whether or not an extractive industry development should go ahead. One such case, in Norway's Kautokeino municipality, is also discussed below.

Any anticipated, yet unrealised, major economic development projects – and the regulatory, stakeholder, business and scientific processes that attend them – can be seen as resulting in 'unbuilt environments' of often invisible effects.[12] In some cases, infrastructure is actually constructed without being used, or is used for only one or two exploration seasons, such as the abandoned oil industry harbour infrastructure built in the Greenlandic village of Aasiaat. After a brief review of methods, we turn to three case studies of such 'unbuilt environments' in the Arctic, in Greenland, Russia and Norway respectively. We seek to explore and identify some of these effects, considering what might be characteristically 'Arctic' about them. We also consider the ways in which our analyses diverge, and how this illustrates the diversity of Arctic experience.

Methods

This chapter draws upon three sets of field work, in Greenland, Russia and Norway. Semi-structured qualitative interviews were a key method in all cases. In Upernavik, Greenland, a total of 16 qualitative interviews were conducted in 2013 and 2014 in Kalaallisut, the Greenlandic Inuit dialect. The research focused on capturing the expectations and aspirations of people living in the area, so as to document and understand their perspectives and the potential for the possible recruitment of locals to work in the industry while also securing local benefits.[13] In Murmansk, Russia, a set of 21 qualitative interviews were carried out (in Russian) in April 2013 with government officials, company representatives, indigenous and civil society representatives and a sampling of 'everyday citizens' who had no direct connection to the oil and gas industry.[14] The research aimed to understand how the urban Arctic residents of Murmansk reacted to and understood an unrealised petroleum development, how they envisioned the future of the region; and how they judged the petroleum companies' corporate social responsibility (CSR) efforts. In Kautokeino, Norway, a total of 26 qualitative interviews were held in 2015 and 2016 with rural residents living close to a proposed mine site. The interviews were held in Saami, Norwegian or English, with translation into Russian or English for the benefit of a multi-national research team. The aim was to understand the extent to which international standards and guidelines on ethical performance in the extractive industries are implemented at the local level.[15]

The case studies

Our case studies are linked primarily by the fact that in all localities a major extractive industry development was actively anticipated by the local community but ultimately did not take place. In the Greenlandic and Russian cases, this was for reasons beyond local control; in the Norwegian case it was a conscious decision made by the local municipality. Two of the case studies involve rural indigenous communities, but the Russian case study is of a non-indigenous urban population in Murmansk – the world's largest Arctic city. The Greenlandic and Russian case studies relate to offshore oil and gas, while the Norwegian case study relates to a proposed gold mine. Our aim is therefore not to draw direct comparisons or make scientifically grounded propositions, but to illustrate a range of local responses to a phenomenon – the unrealised project – that has been covered very little in the academic and policy literature to date.

Upernavik, Greenland

Oil and gas exploration in Greenland has been taking place since the early 1970s without any commercial discoveries yet being made. After a general low level of activity, the beginning of the new millennium brought remarkable increases in the global market price of crude oil (from less than 30 USD/barrel to more than 100 USD/barrel after 2007) and a subsequent increase in the exploration interests of oil companies in Greenland.[16] The Government of Greenland (*Naalakkersuisut*) consequently released a hydrocarbon strategy for Greenland in 2002, announcing new licensing rounds for blocks offshore West Greenland in 2002, 2003 and 2004. In 2008, the US Geological Survey published assessments of large quantities of undiscovered oil and gas resources in the Arctic. The survey indicated that offshore areas between West Greenland and East Canada could hold seven billion barrels of oil, while areas offshore East Greenland were estimated to hold nearly nine billion barrels of oil. The presence of significant gas reserves was also estimated in both offshore areas. *Naalakkersuisut* then released a second hydrocarbon strategy in 2009, which included a new licensing round in North West Greenland in the area of Baffin Bay in 2010 and a two-phased licensing round offshore North East Greenland in 2012 and 2013 (Figure 12.1).[17] The Baffin Bay licensing round led to seven new exploration licences and the licensing round in North East Greenland led to

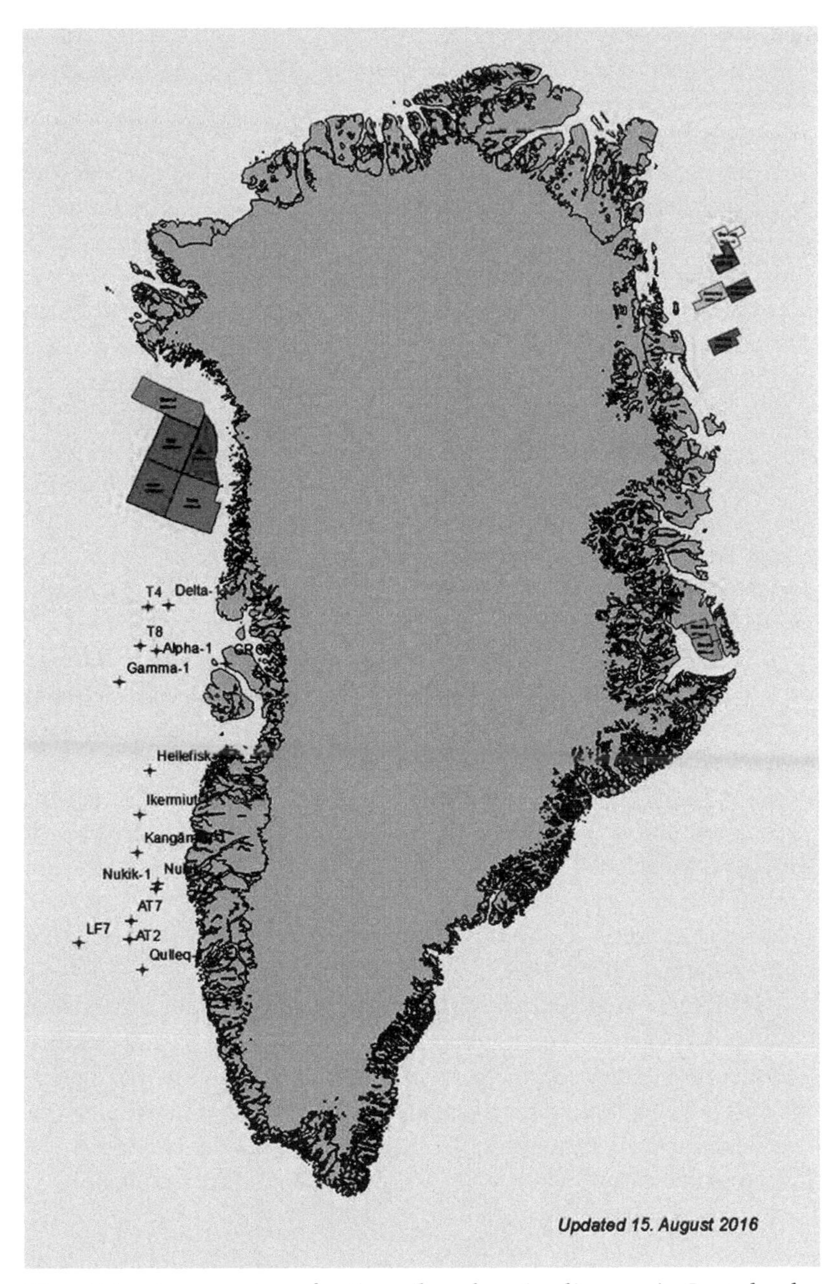

Fig. 12.1 An overview of active oil exploration licences in Greenland (from and used by permission of NunaOil A/S).

four new exploration licences. In 2010 and 2011, Cairn Energy drilled eight wells offshore Central West Greenland. However, all wells were declared commercially dry.[18]

In 2012, a consortium of oil companies with exploration licences in Baffin Bay drilled 11 so-called 'shallow core holes' to evaluate the area. A further four operating companies, including Maersk Oil Kalaallit Nunaat, ConocoPhillips, Cairn Energy PLC and Shell Greenland, held licences to a total number of five blocks in the Baffin Bay area. Seismic exploration and site surveys were undertaken here in 2012 and 2013. The activities were the most extensive in any area of Greenland to date, and all taking place in the sea off Upernavik District. The exploration was expected to lead to the production of oil and related industrial activities; activities that could bring significant change to the communities, both in terms of impacts on nature, the local economy and social structures. During preparation for the exploration programmes, the operating companies were legally requested to, and did, undertake environmental and social baseline studies. They visited and engaged with the local communities to inform them about activities and to manage expectations.

In 2014, the Government of Greenland presented a new strategy on minerals and hydrocarbon resources. This strategy specifies selected areas to be announced for new licensing rounds or open door procedures including the areas of Baffin Bay, Davis Strait, west of Nuuk, Jameson Land, Nuussuaq Peninsula, South Greenland and South West Greenland. But in 2014/2015, the oil price dropped and the level of activity in Greenland similarly declined. After some years of holding on to their licences, several operating companies decided to give them up in 2016. The licences to only ten blocks are still active in 2016, including those in Baffin Bay.

The uncertainty regarding whether industrial development related to oil and gas exploration and extraction will take place in the future and the potential for social change if commercial finds are made place the people living in the Upernavik area in a situation of uncertainty. In the following section, we describe how the exploration activities were perceived at the time, and how the locals coped with the uncertainty.

Local populations and livelihoods

Upernavik District covers 448 km of coastline in North West Greenland. The area includes the town of Upernavik with about 1,100 residents and nine smaller settlements with populations of about 1,700. The inhabitants are, as in most communities in Greenland, predominantly Inuit by ethnicity. The main occupation in the area is hunting and fishing, which is practised both as a commercial and a recreational activity. Families travel to traditional or communally shared hunting, fishing and gathering places along the coast, inlets and smaller islands.[19] Hunting quotas in Greenland regulate the hunting of selected species, but, while some species, such as seal or Arctic cod remain abundant, other animals, such as narwhals and belugas, remain subject to government regulations.[20] Whale quotas are set by *Naalakkersuisut* annually and subsequently distributed to local districts where the municipal authorities decide on the allocation of commercial and leisure hunting licences.[21] Other hunted species include seabirds, walruses, seals and polar bears.[22] Commercial and subsistence fishing, as well as the hunting activities are considered important supplements to the economy for many households.[23]

Local expectations

In 2012 and 2013, when the licence-holding companies were gathering seismic data to map geological features of the sub-surface, a number of public consultations were undertaken and meetings took place between the people of Upernavik District and company representatives.[24] According to the interviews that we subsequently held with people in Upernavik District, they were very aware of the activity and the presence of oil companies and their plans. They did not, however, distinguish between individual companies but rather perceived the industry as 'one' entity. They did not seem very affected by the ongoing activities and in general they expressed relatively little interest in the industry. They did, however, express concerns regarding the potential influence of the activities on their (whale or fish) catch and were also curious to hear more about what kinds of industrial activities were going to take place. They were also curious to hear more from the companies about opportunities to work for or in the industry.

A representative of the municipal office in Upernavik provided the following explanation for why people were interested: 'The word "oil" has been mentioned many times, so there is a general feeling of understanding of oil being important, but the real physical understanding of what oil exploration is and what it means is not clear to people.'[25] Some of the young men in the area expressed an interest in potentially supplementing the income of their hunting activities with jobs in the oil industry in order to support the life they already lived. They expressed a generally positive attitude towards the oil industry, which they saw as a potential facilitator of some of the changes needed locally to uphold their desired way of living in close connection with the land.

There were no high hopes in relation to the activities, but rather a curiosity and an interest from the locals. There were, however, great expectations in Nuuk among the government officials and politicians at the national level. It is also worth noting that expectations were much higher in relation to mining developments and the proposed construction of an aluminium smelter in the southern regions of Greenland, perhaps because these activities are taking place onshore and potentially have a much more direct impact on local livelihoods.[26]

Reflections on the Upernavik case study

In Upernavik, the attitude of the locals and the pragmatic reaction to potential development could be seen as characteristically 'Arctic'. The people of Upernavik were not very influenced by the 'hype' of the potential oil developments. They focused on what was known (birds in the hand and not in the bush) and held on to the importance of traditional activities, rather than dreaming about the future. In a similar way, the Inuit living in the small communities on Alaska's North Slope after 40 years of oil production still have a primary focus on traditional activities rather than on the potential for working in industry or changing or modernising their communities. This also means that local benefits in the Arctic are not necessarily obtained through skills training or the creation of job opportunities in the extractive industry itself (as has been the case in other parts of the world). Stronger and more sustainable communities are achieved instead by securing healthy living standards for people by providing the necessary infrastructure (housing, clean water, transport and supplies), and definitely not from paying out dividend cheques.

Murmansk, Russia

The Shtokman gas field, located in the Barents Sea some 600 km north of the shores of the Kola Peninsula, is one of the world's largest natural gas fields. Development of the field had been discussed in earnest since the mid-1990s. Anticipation on both sides of the Norwegian/Russian border reached fever pitch around and after 2005, when cooperation agreements to develop this field were signed by Russia, Norway and France, with Gazprom at the forefront. This triggered an avalanche of bids for field development[27] and efforts of companies to profile their technical, financial and socially oriented capacities. Eventually Statoil and Total joined a consortium with Gazprom called the Shtokman Development AG in 2008. However, exploration never got off the ground, with the 'shale revolution' in the USA driving gas prices down in what had been a target market for liquefied natural gas from the Shtokman field.[28]

Great expectations

The interview findings were illustrative of the effects of extractive anticipation in two key regards – negative views on future prospects and changed understandings of potential extractive stakeholders.[29] There are also, potentially, a myriad ways in which the Shtokman development may have had lasting impacts on the region outside of the ones identified via the interview set. Should the case study site have been located in Teriberka on the Murman coast, the impact of actual physical changes in the environment, including advanced infrastructure and changed budgeting or infrastructure planning, may have been more evident. One may also have found more individual decision making directly influenced by the prospect of the project (building choices, business plans, educational decisions and so on). As the case study presented here was focused on tracing the broader regional impacts of the Shtokman project, in particular the expectations and recollections in the regional capital city of Murmansk, the impacts of anticipation remain more cognitive and collective rather than individual or material.

First, there was an impact on how respondents perceived the economic prospects of the region. Some argued that oil and gas had remained entirely 'virtual', yet had still managed to have a negative impact on the region. Local respondents recalled overly optimistic personal spending and borrowing in the course of the build up to the expected project. Interviewees from business, NGOs and the public sector argued that housing prices had become inflated during the days of Shtokman mania, but not matched by employment and salary growth: 'Just say the word Shtokman and apartment prices go up', was a comment that several interviewees made.

Second, respondents' experience with and perceptions of the petroleum companies' corporate social responsibility (CSR) efforts seem to have catalysed change in some understandings among the broader web of stakeholders in the region. In other words, the Shtokman project likely changed perceptions of who can and should play a role in shaping major new extractive projects in the region and we should consider how those new understandings may play a role in future prospects as well.

On the whole, respondents had fairly strong recollections of the social policies and efforts of the international oil companies (e.g. Rosneft, Total and Statoil) that had vied for a position in the Shtokman project. Many interviewees had positive memories of international petroleum companies' advance engagement in the region, such as support for business alliances and NGOs, youth engagement and musical and cultural events.

When it came to the environment and also the capacity of companies to bring sustained long-term benefits to the region, however, several interviewees had developed a more sceptical understanding of 'new' extractive actors (even though their operations never reached the stage where these benefits or risks materialised). Other respondents were uncertain about the actual outcomes of CSR, wondering if it had been just PR or empty words to satisfy company policy. One interviewee from the public sector put it this way: 'We have CSR on paper only. I wish companies understood that they have a responsibility not only to their managers and owners – but to all of us who live here.'

Whether the interviewees had negative, positive or neutral recollections of these concrete CSR practices, for nearly all the interviewees, the memories and current perceptions of the major economic actors that established themselves in the region during the Soviet period were an important conceptual touchstone. These longstanding industrial actors (mining and metallurgy, shipbuilding, nuclear power plants) were held up as the standard against which the social performance of the 'newcomer' petroleum companies (both domestic and international) was judged. Interviewees from all walks of life warmly recited past and present benefits and services provided by the companies to their own employees – entertainment and celebrations, travel, pensioner housing, specialised medical care, education and other family benefits. This renewed appreciation for existing industry can be seen as a lasting imprint of the Shtokman project and may be important in steering regional politics. How will these companies be treated in the future? How hard will they be pressed (or not) by regional government or the public on social and environmental issues as they arise?

In light of experience from the unrealised Shtokman project, the regional authorities interviewed described themselves as limited in holding any large economic actor to account in social and environmental matters. The possibility that companies can 're-register' their tax home to another region was mentioned by three regional government interviewees as causing them to focus on providing 'hospitality', 'maximum comfort' and 'being appealing' for business, rather than pushing for high social and environmental standards. Regional authorities saw their role in relation to oil and gas companies as especially problematic. In the words of one involved regional civil servant:

> The development of this sector is carried out by companies of federal significance. Because of this, many of the strategic decisions about them are taken in Moscow. But there is a huge number of tasks that need to be carried out by regional and local levels, we have to create conditions for building of commercial objects, infrastructure, roads ... not least the right social conditions. We know the region best and a lot of these tasks can be carried out by us more efficiently and quickly.

Environmental organisations also saw themselves as important participants in shaping industrial development in the region and overall felt that their engagement with companies had been constructive, even while they maintained a vigilant attitude toward the companies involved. They felt increasingly well-educated by the process of engaging with petroleum companies new to the region, including being further attuned to the international practices and standards that may serve as pressure points on companies (as they are important for companies' access to international finance).

One interview was conducted with a representative of an indigenous Saami organisation visiting Murmansk. Here, the geographical focus on Murmansk city is limiting as most Saami organisations representing the approximately 2,000 Saami people in the region are headquartered elsewhere on the Kola Peninsula. This interviewee painted a worrying picture, arguing that the Saami had not been effectively consulted in terms of commercial developments and saying they felt they had been affected by industrial development relating to the offshore in subtle ways but that it is hard for them to prove causal links (with the burden of causality left placed on them). Other interviewees were dismissive about indigenous interest groups, indicating a divide in public engagement vis-à-vis the Shtokmann project along ethnic lines.

Reflections on the Murmansk case study

Despite the project remaining unrealised, a cross-section of the public in Murmansk nevertheless possessed well-developed expectations and perceptions of the oil and gas companies that had jockeyed for positions around the Shtokman gas field. Interviewees were reluctant to engage with the concept of CSR that the companies had brought with them and instead referred warmly to a gold standard of past and present employee benefits set by the industrial complexes of the Soviet period. Interviewees also had clear perceptions of their own and others' potential roles as stakeholders in managing an oil and gas future that had not come into existence. These findings suggest that the anticipatory practices around the Shtokman field have had lasting repercussions for how economic development and environmental risks are understood in the region and for shaping understandings of what kinds of stakeholders matter for large-scale economic development.

Kautokeino, Norway

In Norway, uncertainty around offshore oil and gas development has influenced national government efforts to revive its mining sector, which is focused particularly in Finnmark County in Northern Norway, where Saami reindeer herding is most intensely practised. Norway has opened no new mines in 30 years; therefore recent developments in Kautokeino and neighbouring Kvalsund have attracted great interest.[30] While the Kvalsund copper mine may go ahead, the decision by Kautokeino to refuse a proposed gold mine has caused shock and questioning within Norway.

In September 2015 Kautokeino's municipal council placed a four-year moratorium on discussions about whether or not to re-open their existing copper/gold mine, known as Biedjovaggi. The municipality had twice rejected proposals by Swedish mining company Arctic Gold. Municipal leaders argued that reindeer herding is critically important for local livelihoods and the Saami culture, and they would prefer to protect and support the reindeer herding families who make up over half of Kautokeino's population of 1,386.[31] Mining is not the only threat to herding, although a map of exploration licences in Finnmark County reveals a land scattered with claims. It is one of many (cumulative) threats, including wind farms, roads, electric power lines, tourist cabins and hydropower projects.

Kautokeino had experience of mining from the 1970s to the early 1990s, when the previous copper/gold mine was closed (for the second time).[32] Today, Kautokeino has a budget deficit and one of the highest unemployment rates in Norway at 6.4 per cent, almost twice the national average of 3.3 per cent.[33] Reindeer herding is the largest economic activity in Kautokeino, but it cannot provide for everyone.

The Kautokeino decision: how and why did the project not happen?

Finnmark County has a special status supporting the rights of the indigenous Saami, who make up around 10 per cent of the total population. Kautokeino municipality is situated in inner Finnmark, which has historically preserved traditional livelihoods and Saami language more than the coastal regions of Finnmark and has the largest concentration of reindeer herders in Norway.[34] It is one of only two municipalities where the majority of the population is Saami and where the Saami language is used by most people in daily life. The practice of reindeer herding is important for maintaining the language and is protected through the Reindeer Act (2007). Saami rights are also protected by legislative developments in the 1980s and 1990s and the establishment of the Saami Parliament in 1987.

Arctic Gold took ten years to obtain an exploration licence for the Biedjovaggi mine from the Norwegian government. They succeeded in 2011 and invested heavily in exploratory drilling. The proposal was for an open pit mine, greatly expanding the footprint of the existing mine on land currently used as reindeer pasture. A clause in Norway's revised Planning and Building Act (2009) allows municipal councils to decide whether or not to move forward with a mining project at the stage of environmental impact assessment (EIA). In April 2012 Kautokeino's 19-member municipal council, with a narrow 10-9 majority, voted not to allow Arctic Gold to do an EIA. Those who voted against the mine argued that people were well aware of the impacts of mining from previous experience and did not need an EIA. A further concern was the fact that a decision made after the EIA was completed could be challenged at the ministerial level in Oslo, thus taking power away from the municipality.

Following the 2012 vote, Arctic Gold challenged the legality of that decision; offered to carry out a social impact assessment (not mandatory according to Norwegian law); and excluded the southern part of the proposed mining area, which was most important for herding. They drafted an agreement with the municipality to support local business, culture and infrastructure. Arctic Gold's CEO also offered a one-off payment of NOK 20 million and stated that a further 'no' would mean that Norway's mining legislation was not working properly. He was labelled 'arrogant' in the press: a picture of him in a Texan hat was circulated on social media and he was dubbed 'the cowboy'. A second refusal came in December 2013, with a similar narrow majority (10–9). Company representatives admitted they had not realised the importance of reindeer herding.[35]

Arctic Gold indicated that a further attempt was possible after Kautokeino's municipal elections in September 2015. However, the new council announced immediately that there would be no further discussion about the mine for the rest of their four-year term in office, as they wanted to focus on other things. The current moratorium is not a definitive 'no' and much remains to be done if the conflict is not going to emerge again.

Local perceptions and responses

One of the most striking observations from Kautokeino was the powerful effect that the mere prospect of the mine had on the community. It exacerbated tensions along existing fault lines, with non-herding Saami claiming that the herders did not want the rest of the community to develop and revealing resentment at the legal rights that have been afforded the reindeer herders to date. A strong supporting voice for the mine came from the political party that was established to defend the interests of non-reindeer herding Saami following the enhancement of legal rights for herders. Views were not always clear cut, however, and pro- and contra-groups were also deeply intertwined through family and communal ties.

Another striking observation was the contrast in different ways of imagining the future, between the state and the community, and within the community itself. For instance, researchers at Kautokeino's Saami University College have explored the chasm between the state vision of future resource development, based on grand economic projects underpinned by science and technology, versus the longer-term and historically rooted vision of the herders, based on customary practice, adaptive management and collective use of land, rather than private ownership.[36]

One positive reason for supporting the mine proposal was the potential for enlivening the local economy. Local residents remembered the previous period when the mine was open. At that time there was also a military camp, and mine workers and soldiers visited the community, people went out more, spent more money and the community was livelier. People also remembered that wages at the mine were higher than in other places locally.

A few local businesses would directly benefit from the re-opening of the mine, for example those that provide drilling services or specialist machinery. Kautokeino is heavily reliant on public sector employment – which is reportedly 70–75 per cent of total employment (compared to the town of Alta to the north, which has more commerce and where only 30 per cent of jobs are in the public sector). In Kautokeino, there are few shops and people regularly go shopping in Alta – a three-hour bus journey away – but are reluctant to set up their own shops. For some, the mine would be the answer to economic stagnation in the village.

One business respondent, however, observed that all the talk of the mine was draining positive energy and enterprise potential from the community:

> It has such as psychological impact. The thought that this could be the solution. It's like a grey cloud. Because young people want to stay. They want to go and get an education and then come back and use it. There is huge potential. We kill this potential with the mine question. I'm afraid of this more than the mining itself.[37]

In 2015, a local official closely involved with the municipal council decision pointed out that of Norway's 428 municipalities only a few have mines: 'The illusion that a municipality has to have a mine or it dies is not true. We can benefit from a mine but there are other opportunities.'[38] He emphasised the sustainability of the reindeer herding industry in Kautokeino and the fact that it is a large part of the reason why young people want to stay in the municipality. The official stated that the mine decision was primarily about Saami responsibility for traditional lands, and the need to respect international indigenous rights. Norway has ratified the International Labour Organisation Convention No.169 on Indigenous and Tribal Peoples (1989); and supports the UN Declaration on the Rights of Indigenous Peoples (2007), both of which require local level decision making by indigenous communities relating to resource extraction projects. The official said there would be no further negotiation with extractive companies unless there is dialogue between those companies and the rights holders themselves. There has also been talk of setting aside the land permanently for reindeer pasture, but he observed that this would not be supported by many in the local community and would be a very complex process.[39]

Reflections on the Kautokeino case study

Despite support for indigenous rights in Norwegian legislation and institutions, the 'extractivist' economic model promoted by the state is at odds with herders' own vision of the future. Progress in indigenous rights legislation moreover appears to have caused resentment in a mixed community where reindeer herders are perceived to benefit more from government support than non-herding Saami. For those who want the mine, it is seen as a 'saviour' project that will address problems that could be addressed in other ways, although these alternatives are poorly understood as yet. The municipality has much to do over the next four years to resolve some of these internal community issues and ensure that the land users are given adequate representation at the decision-making table, while others who feel disempowered or disadvantaged are also allowed the opportunity to have their views incorporated into future planning processes.

Concluding discussion

As Arctic cooperation continues to expand, most recently via the newly established Arctic Economic Council, attention to the limits of shared ideas and practices and the abiding significance of realised and unrealised local developments remains essential. Otherwise, it may be difficult to understand the dispositions, policy trajectories, political processes and expectations that Arctic residents bring to future debates about circumpolar economic and social development. One conclusion we have reached in the course of our research around these case studies is that, as yet, the phenomenon of the unrealised project has been covered very little in academic and policy literature, unlike the notion of 'boom and bust' for instance.

This discussion explores the findings of our case studies from two angles:

- What do our cases tell us about projects that have not happened?
- What do they tell us about Arcticness?

What do our cases tell us about unrealised projects?

The case studies presented in this chapter have illustrated some of the ways that extractive industry development (often assumed to be an unstoppable force) is by no means a guaranteed outcome, even where ambitious plans are in place and anticipatory actions well underway. Shtokman remains an unrealised oil and gas 'megaproject' and Greenland's oil and gas industry has yet to get off the ground, while Kautokeino municipality has taken the opportunity to reject a proposed project that could undermine traditional lifestyles. Despite the fact that these projects have not gone ahead, all of the cases demonstrate the extent of local impacts from a development, even before it has actually started – something that is rarely taken into account in the analysis of industrial impacts on local communities.

The case studies have yielded some ideas about the different stakeholders who can influence these processes and their capacities. This influence can happen at different levels, with the tension between 'centre' and 'municipal' levels evident in all cases. In Murmansk, the offshore developments also heightened local expectations about corporate responsibility practices, with long-established industrial entities comparing favourably with the 'newcomers'. The study also revealed the importance of considering how local, international and regional standards and discourses brought in by media, international companies and circumpolar cross-border interactions combine to shape 'unbuilt landscapes' in novel ways.

The cases revealed that local communities are far from homogenous, even if all the local residents are from the same indigenous ethnic group, and the prospect of a new development can open up existing internal 'fault lines' within a community. For example, in Kautokeino, latent resentment about the benefits that reindeer herders receive from the state was intensified when non-herders perceived them as seeking to halt a potential alternative economic option for the community.

While in some cases, such as the Upernavik case, local people are not fired up by the 'hype' of a new project, in other cases, local hopes for profits from the extractive industries can be so intense that they crowd out the potential offered by other socio-economic development paths, as was the case for some residents of Kautokeino. The expectation of extractive industries may result in anticipatory activities such as the Government of Greenland developing and updating its hydrocarbon strategy; or the artificial inflation of the housing market in Murmansk. And yet external forces might suddenly undermine development prospects, leaving communities struggling to revert back to more self-sufficient modes of development.

What do our cases tell us about Arcticness?

Our case studies suggest a number of factors that could point towards an understanding of the notion of 'Arcticness'. Sometimes these are better seen in terms of a cluster of factors that might come together uniquely in the Arctic; sometimes these are striking similarities that can be perceived in different parts of the Arctic, although not in all communities throughout the region. Two of our case studies focus on small indigenous communities in isolated localities practising traditional livelihood activities and, perhaps, this is a dominant picture that many outsiders have of the Arctic. Yet the Murmansk case study highlights the fact that there are also city populations living north of the Arctic Circle. Meanwhile, the Kautokeino case study illustrates the challenges of modernisation in an indigenous community when only half the community practises traditional livelihood activities (subsidised by the state).

Extractive industry development can threaten people's connection to the land – something that deeply defines existence for many Arctic residents, particularly those from indigenous communities. Some people seek to keep their ties to their land and resource-use practices strong. For example, the Upernavik communities hold on to the importance and value of traditional activities, rather than dreaming about the future and the possible benefits that externally-imposed modernisation might bring them. This can be compared to observations of Inuit practices in Alaska, despite 40 years of oil production. It is also comparable to the vision of the reindeer herding community of Kautokeino, whose vision of the future contrasts with the extractivist economic model promoted by the state. Yet half of the Kautokeino community, like others across the Arctic, still sees extractive industries as the easy answer to a multitude of local issues, including youth unemployment, economic stagnation and the maintenance of local infrastructure and public services.

Arctic communities may be disproportionately exposed to the experience of unrealised extractive futures. This is often due to events that local people have had little control over, such as commodity price fluctuations, which may lead companies to withdraw from the Arctic first of all as it is one of the most expensive places to work. Sometimes a project may be halted as a result of local voices making themselves heard in deciding against a development. Yet despite great advances in the understanding and defence of indigenous rights, this is uncommon in the Arctic. Our case studies portray the range of opportunities from mineral resource development, as well as the depth of uncertainty surrounding every development, and the way that decisions, once made, may be thrown up in the air with a turn in commodity prices, or a new municipal election. As such, Arcticness might partially be defined by the regular experience of ambitious, single-industry plans for development, some of which come about and many that do not, but all of which leave their traces.

A key task for companies and policy-makers promoting their visions of the future is to communicate the fundamental uncertainties involved in realising them, and discussing the ways in which anticipation is not the same as prediction or certainty. For researchers and policy analysts, there is a need to explore further the issues surrounding the 'unrealised project', including analysis of project impacts that take place before a project is confirmed (such as anxiety, community tension, unrealistic or heightened expectations, and the 'crowding out' of other future options); the different factors that may result in a project not being pursued, including issues ranging from commodity price fluctuations to the different ways that communities are able to 'say no' to a project; and the range of different outcomes that might follow, be it economic decline or the emergence of local enterprise.

13
Editorial Conclusion: Arcticness by any other name

Ilan Kelman

Does Arcticness convey power and voice from the north? Does it disempower, with an alien concept and artificial construct foisted on diverse peoples and regions who have little in common apart from living at high latitudes?

The chapters within this book open this conversation, dissect the concepts, put forth numerous queries and provide few answers. They do provide pathways towards responding plus indications of the variety of answers which exist. In problematising and de-problematising both the Arctic and Arcticness, they promote and dispute views from inside and outside the region, embracing and challenging multiple definitions.

Consequently, the diversity of the Arctic and of Arcticness emerges. Diversity which, perhaps, is so wide-ranging as to deny any graspable description or characteristics of what referring to the Arctic really means. Yet an undeniable materiality of high latitudes produces an environment – including climate, geology, ecosystems, biota, air, water, land and ice – shaping life and livelihoods differently than environments at lower latitudes.

While also displaying similarities. The comparative chapters illustrate this. High altitudes sport similarities with its anagram of high latitudes, as demonstrated by Tibet. Small, resource-dependent communities produce analogues between Greenland and Uganda. Other topics remain unexplored, such as high-latitude communities in the southern hemisphere, although those extend to a mere 55°S, far in distance and concept from many definitions of polar areas.

One consequence is that northern high-latitude communities are unique on Earth. Mountain communities might have similar climates, but they do not enjoy the day–night imbalance witnessed throughout Arctic summers and winters. Nor do mountain communities (by definition) necessarily sit at low elevations, including sea level, as numerous Arctic communities do.

This level of similarity around the northern latitudes cannot by itself define a region. The multiple definitions of Arctic attest to the peoples being more than their environments.

It is not even clear that all peoples living at high northern latitudes share an assumption of being Arcticly similar. Many certainly do, entirely embracing the Arctic concept and seeking out those at matching latitudes. Others accept Arctic similarities without presuming those to entail uniqueness from non-Arctic peoples. Many reject both premises, challenging the importance or differentiation of the Arctic from elsewhere.

No view is especially right or wrong. People are entitled to their perspectives and to act according to their own Arctic-related wishes. Then, what does it mean for Arcticness and what does Arcticness mean?

This volume's contributions and contributors answer with mixed results on a solid baseline through a series of contrasts which are complementary rather than contradictory. Notions of Arcticness are material and emotional; products and processes; manufactured externally and coming from the people as who they are; an innate and prevalent trait of living in the region; and an assemblage according to what the assembler desires.

A danger coalesces of Arcticness being everything and nothing. It represents exceptionality and uniqueness – just like everyone and everywhere else!!

Nevertheless, these apparent incongruities do not obviate the need, desire or utility of Arcticness. Arcticness permits expression of what is felt and seen from being in and from the Arctic, alongside what is felt and seen externally. It both gives and takes both power and voice.

It gives, as Hansen writes, through 'interconnectedness of characteristics and perspectives on quality of life' summed up by Medby as the 'quality of being Arctic' which she then further interrogates. It takes through Naess' expostulation of 'exotification' and through Tilling and colleagues' concern about the scientification of the Arctic, which can never quite capture the connection they seek in their scientific work.

How do we avoid the Arctic being simply, in Duda's words, 'featured in outsiders' collective constructs', territorialised as per the chapter by French and colleagues, or commodified as detailed by Wilson and colleagues? Why do we even wish to avoid these phenomena?

Ultimately, as with any concept or idea, it is what is made of it, to be used, abused and misused as those with power choose. Arcticness lends itself to adages applied to so many other phenomena and processes, as ephemeral and operational, as theoretical and grounded, as entrenched and severed. Reminiscent of democracy, Arcticness could be the region's worst descriptor – apart from all the others. Pilfering from participatory development discourse, the act of labelling with and as Arcticness could be an imposed tyranny. McCauley and colleagues propose Arcticness as a process rather than as a product – exactly as was done with post-disaster shelter two generations previously.

Arcticness is thus made to have traction and relevance beyond the Arctic, often by those seeking to appropriate and misappropriate the Arctic for their own purposes. The starkest example is the polar bear being conscripted to symbolise Arctic change, neglecting the peoples, livelihoods and communities who are affected by change far more than polar bears. Ice, snow and cold are important Arctic symbols, but not for their own sake – instead, for the peoples' Arcticness, epitomised in Sheila Watt-Cloutier's powerful statement about 'The Right to be Cold'.

This power and voice from the north, defining themselves by themselves, is perhaps the key of and for Arcticness: how the Arctic is lived and experienced by peoples and communities along with the qualities therein. The chapters in this book are experiential, beyond memory, meaning that by definition much is missing and much is disputable, because people experience and articulate differently.

Here, perhaps, lies the meaning of Arcticness beyond the Arctic: to take control of one's own descriptors and explanations of oneself, to create and express power and voice by one's own definitions of oneself, and to grasp and tackle the challenge of others setting the agenda for oneself – all happening inside and outside one's own community. Opinions will differ. It can and should source strength rather than battle.

The same has been witnessed for islandness, ethnicness and engineeringness among many other -ness-es. These reflections seek identities, qualities, groupings, connections and boundedness. The Arctic is no different, including with respect to the tumultuous environmental and social changes of contemporary times – and reaching back through millennia. Arcticness might seem to cleave by setting the Arctic apart, but in distinguishing what is or could be Arctic from what is not, a form of 'Arctic without borders' – borders across time and space – is sculpted.

Connections among indigenous peoples, natural resource-based communities, cold weather locations, mining towns and ocean-based livelihoods have all contributed to Arcticness. In otherising what is not

Arctic, bonds are forged through identifying differences which by definition delineate what are not differences. This inevitably bounces back to those being otherised who must similarly describe themselves partly through what they are and partly through what they are not.

Arcticness thus teaches how to define and accept one's own -ness as distinct and partitioned from others, yet also with similarities and boundary crossings. It is not even about iconising. Instead, it is about taking the power and creating the voice in, for and of the Arctic which for too long has resided outside the northern latitudes – to a large extent revealing an 'Arctic of the oppressed'.

It is about creating an Arctic home for those living in this ever-changing home.

Whether or not 'Arcticness' is the most powerful and voiceful term remains to be seen. However it is labelled, and preferably with transferability across Arctic languages and cultures, Arctic experiences and qualities resonate far beyond the location and, from the authors writing here, it proffers knowledges, wisdoms and actions to the world for addressing today's global challenges and opportunities.

Afterword
Within Arcticness, outside the Arctic

Vladimir Vasiliev

What is the Arctic and what is Arcticness for a person living in the largest Arctic region of the world, not living in the Arctic zone itself, but much more southwards, and still considering himself to be an Arctic resident?

I am a native Sakha. I was born and raised in Central Yakutia, in taiga area. I first saw tundra and the Arctic Ocean being an adult 27-year-old postgraduate student at the Yakutsk Scientific Center of the Siberian Division of the Russian Academy of Sciences. At the same time, from an early age, I was absorbed by the stories about brave polar explorers in an effort to understand what attracted people to such a harsh land, why, from olden times, they have so wilfully struggled through snow and cold into this, at first sight, deserted country, knowing that they may not come back.

I can confirm the words of all people who fall in love with the Arctic at first sight, that it enchants and keeps attracting you. Vast expanses, bright colours of tundra in summer, white land melting into the sky in winter, incredible glows of the northern lights, reindeer herds of many thousands, somewhat unreal and the absolutely ancient life of indigenous people in the reindeer-skin tents which seem to take you back thousands of years. Even the presence of some evidence of modern life such as newspapers, books, walkie-talkies and televisions do not hamper perceiving the Arctic as a separate ancient world, as another civilisation.

At the same time, the Russian Arctic has had its periods of prosperity. In Soviet times, the Northern Sea Route functioned at full capacity and large-scale industrial projects of tin and gold mining were launched in the Yakut Arctic. The Arctic regions had much better supplies of all kinds of commodities than those in Central Yakutia. The young willingly went to work in the remote settlements, being confident that everything necessary would be provided for proper work in the Arctic.

The situation changed dramatically after the collapse of the Soviet Union. The development of the Arctic territories almost stopped. Only a few ships passed along the Northern Sea Route in summer, many settlements were closed, people were leaving, and the only ones to stay were the indigenous people and those who did not see themselves living in other regions of the country, whose hearts had been chained by this harsh land.

Almost a half of Yakutia was in a very difficult situation. Immense distances and lack of funds in budgets did not allow the solving of all the problems at once, but the leaders of Yakutia have always paid great attention to the development of the Arctic zone. It was obvious from the very beginning that a single region was not able to tackle all the problems, and the first President of the Sakha Republic (Yakutia), Mikhail Nikolaev, actively established connections not only with Russian regions, but also with the foreign ones, in order to attract global attention to the development of the Arctic as a whole. For this purpose, he initiated the republic's joining the Northern Forum international organisation and supported the establishment of the University of the Arctic and other international structures.

Being a biologist, I wanted to make a contribution to environmental protection in the Arctic. It is now pleasant to remember that I was among the initiators and coordinators of the Integrated Arctic Expedition of the Yakutsk Scientific Center of the Siberian Division of the Russian Academy of Sciences in the 1990s. Within the frameworks of the expedition, we were able to collect a considerable amount of data and information on not only biodiversity and ecosystem functioning, but also on subsurface use, the status of indigenous peoples, traditional economies, the preservation of languages and culture. I hope that these materials were helpful in developing the republic's new legislation and development programmes of different regions.

The experience I have received within the framework of the Arctic expedition still helps me. With my colleagues from not only Yakutia, but also from Russia and abroad, we have implemented a whole range of projects on environmental protection in the Arctic, support for indigenous peoples and climate change, involving the capacities of the Arctic Council, Northern Forum, UNEP, UNDP, WWF, Snowchange and other recognised international organisations.

I believe that I have every reason to consider myself a resident and patriot of the Arctic, as I have devoted more than 30 years of my life to studying it and attracting support in different areas.

My family is also closely connected to the Arctic. My wife, Maria Krivtsova, helps me in implementing all projects, not only as an interpreter, but also as a coordinator. She is Russian born in Yakutsk but she is engaged in many international activities dedicated to the Arctic. Even our five-year-old daughter Sofia, although she has not been to the Arctic yet, has her own small reindeer herd of four reindeer – on the day she was born, Turvaurgin Chukchi community gave her a female reindeer. We have become good friends with the members of this and other communities of the Lower Kolyma, as well as with the College of Northern Peoples. Through them, we continue our strong ties with the Arctic.

At present, being a member of the Government of the Sakha Republic (Yakutia), I have many more opportunities for direct participation in the development of the Yakut Arctic, and I intend to do my best to change the life in the Arctic for the better. In Yakutia, 2014 was announced as the Year of the Arctic through the initiative of the Head of our Republic, Egor Borsiov. Such endeavours allow focusing on a specific topic, to evaluate the situation, to conduct high-quality analyses and to provide an integrated approach involving all available agencies. The Year of the Arctic resulted in the development of the Integrated Arctic Territories Development Program and the establishment of the State Committee for the Arctic Issues.

Today, life in the Arctic regions is changing rapidly. New community facilities – schools, kindergartens, hospitals, cultural and sports centres – are commissioned. Large-scale work on developing industry, attracting investments and improving the entire energy supply system is conducted. In 2015, the most powerful (1 MW) solar power plant above the Arctic Circle was constructed in Batagai settlement of Verkhoyansk region. Constructing a wind plant of the same capacity is planned in Tiksi. The population of the Arctic regions is growing slowly but steadily.

It is due to adopting a whole range of laws on supporting the population, especially indigenous peoples, on the transition of planning and implementation of obligations to a programme-based method, as well as considerable improvement of medical services, the appearance of a telemedicine network, and improvement of communications systems (97 per cent of the republic's area is now covered by cell communication and the internet). We see the revitalisation of the Northern Sea Route, a key significance in the development of the Arctic. That is why the Yakut

Arctic is not an economically depressed region anymore. It is a region whose population is optimistic about the future, in spite of the difficult global economic situation.

Work in the Arctic has changed my entire life. It has filled my life with dramatic events and vivid impressions. It has made me friends with lots of people from all over the world. I hope to be useful to the Arctic until the end of my life which, to me, is what Arcticness is about.

Notes

Chapter 4

1 F. Fetterer, K. Knowles, W. Meier and M. Savoie, 'Sea Ice Index' (National Snow and Ice Data Center – Digital Media, Boulder, Colorado, USA, 2002, updated daily).

2 C. Süsskind, 'Who invented radar?', *Endeavour* **9** (2), 92–6 (2015).

3 A. J. Butrica, *To See the Unseen: A History of Planetary Radar Astronomy* (NASA History Office, Washington, DC, 1996).

4 M. I. Skolnik, *Introduction to Radar Systems* (McGraw-Hill, New York, 2001).

5 M. E. R. Walford, 'Radio echo sounding through an ice shelf', *Nature* **204** (4956), 317–19 (1964); S. Evans and B. M. E. Smith, 'A radio echo equipment for depth sounding in polar ice sheets', *Journal of Scientific Instruments* **2** (2), 131–6 (1969).

6 R. W. Jacobel and S. K. Anderson, 'Interpretation of radio-echo returns from internal water bodies in Variegated Glacier, Alaska, USA', *Journal of Glaciology* **33** (115), 319–23 (1987); R. Drews, O. Eisen, D. Steinhage, I. Weikusat, S. Kipfstuhl and F. Wilhelms, 'Potential mechanisms for anisotropy in ice-penetrating radar data', *Journal of Glaciology* **58** (209), 613–24 (2012).

7 P. Femenias, F. Remy, R. Raizonville and J. F. Minster, 'Analysis of satellite-altimeter height measurements above continental ice sheets', *Journal of Glaciology* **39** (133), 591–600 (1993).

8 C. Wunsch and D. Stammer, 'Satellite altimetry, the marine geoid, and the oceanic general circulation', *Annual Review of Earth and Planetary Sciences* **26**, 219–53 (1998).

9 C. G. Rapley, H. Griffiths, V. A. Squire, M. Lefebvre, A. R. Birks, A. Brenner, C. Brossier, L. D. Clifford, A. P. R. Cooper, A. M. Cowan, D. J. Drewry, M. R. Gorman, H. E. Huckle, P. A. Lamb, T. V. Martin, N. F. McIntyre, K. Milne, E. Novotny, G. E. Peckham, C. Schogounn, R. F. Scott, R. H. Thomas and J. F. Vesecky, *A study of satellite radar altimeter operation over ice-covered surfaces; ESA contract no. 5182/82/f/CG(SC)* (ESA Scientific and Technical Publication Branch ESTEC Noordwijk, Holland, 1983); S. Laxon, 'Sea ice altimeter processing scheme at the EODC', *International Journal of Remote Sensing* **15** (4), 915–24 (1994).

10 A. Shepherd, E. R. Ivins, A. Geruo, V. R. Barletta, M. J. Bentley, S. Bettadpur, K. H. Briggs, D. H. Bromwich, R. Forsberg, N. Galin, M. Horwath, S. Jacobs, I. Joughin, M. A. King, J. T. M. Lenaerts, J. Li, S. R. M. Ligtenberg, A. Luckman, S. B. Luthcke, M. McMillan, R. Meister, G. Milne, J. Mouginot, A. Muir, J. P. Nicolas, J. Paden, A. J. Payne, H. Pritchard, E. Rignot, H. Rott, L. S. Sorensen, T. A. Scambos, B. Scheuchl, E. J. O. Schrama, B. Smith, A. V. Sundal, J. H. van Angelen, W. J. van de Berg, M. R. van den Broeke, D. G. Vaughan, I. Velicogna, J. Wahr, P. L. Whitehouse, D. J. Wingham, D. Yi, D. Young and H. J. Zwally, 'A reconciled estimate of ice-sheet mass balance', *Science* **338** (6111), 1183–9 (2012).

11 *Nature Geoscience* **5**, 194–97 (2012).

12 S. Laxon, K. A. Giles, A. Ridout, D. J. Wingham, R. C. Willatt, R. Cullen, R. Kwok, A. Schweiger, J. L. Zhang, C. Haas, S. Hendricks, Ri. Krishfield, N. Kurtz, S. Farrell and M. Davidson, 'CryoSat-2 estimates of Arctic sea ice thickness and volume', *Geophysical Research Letters* **40** (4), 732–7 (2013); R. L. Tilling, A. Ridout, A. Shepherd and D. J. Wingham, 'Increased Arctic sea ice volume after anomalously low melting in 2013', *Nature Geoscience* **8**, 643–6 (2015).

13 J. O. Sewall and L. C. Sloan, 'Disappearing Arctic sea ice reduces available water in the American west', *Geophysical Research Letters* **31** (6), L06209-06201–L06209-06204 (2004);

J. S. Singarayer, J. L. Bamber and P. J. Valdes, 'Twenty-first-century climate impacts from a declining Arctic sea ice cover', *Journal of Climate* **19** (7), 1109–25 (2006).

14 J. A. Francis and N. Skific, 'Evidence linking rapid Arctic warming to mid-latitude weather patterns', *Philosophical Transactions of the Royal Society A: Mathematical Physical and Engineering Sciences* **373** (2045) (2015); J. A. Francis and S. Vavrus, 'Evidence linking Arctic amplification to extreme weather in mid-latitudes', *Geophysical Research Letters* **39** (6), L06801-06801—L06801-06806 (2012).

15 D. J. Wingham, C. R. Francis, S. Baker, C. Bouzinac, D. Brockley, R. Cullen, P. de Chateau-Thierry, S. W. Laxon, U. Mallow, C. Mavrocordatos, L. Phalippou, G. Ratier, L. Rey, F. Rostan, P. Viau and D. W. Wallis, 'CryoSat: A mission to determine the fluctuations in Earth's land and marine ice fields', in R. P. Singh and M. A. Shea (eds), *Natural Hazards and Oceanographic Processes from Satellite Data* (2006), Vol. 37, pp. 841–71.

16 P. Kanagaratnam, S. P. Gogineni, V. Ramasami and D. Braaten, 'A wideband radar for high-resolution mapping of near-surface internal layers in glacial ice', *IEEE Transactions on Geoscience and Remote Sensing* **42** (3), 483–90 (2004).

17 D. F. Page and R. O. Ramseier, 'Application of radar techniques to ice and snow studies', *Journal of Glaciology* **15** (73), 171–91 (1975).

18 J. W. Wood, C. J. Oliver, I. P. Finley and R. G. White, 'Synthetic aperture radar', *Patent US* 4963877 A, (1990).

19 S. L. Farrell, N. Kurtz, L. N. Connor, B. C. Elder, C. J. Leuschen, T. Markus, D. C. McAdoo, B. Panzer, J. Richter-Menge and J. G. Sonntag, 'A first assessment of IceBridge snow and ice thickness data over Arctic sea ice', *IEEE Transactions on Geoscience and Remote Sensing* **50** (6), 2098–111 (2012); J. L. Li, J. Paden, C. Leuschen, F. Rodriguez-Morales, R. D. Hale, E. J. Arnold, R. Crowe, D. Gomez-Garcia and P. Gogineni, 'High-altitude radar measurements of ice thickness over the Antarctic and Greenland ice sheets as a part of Operation IceBridge', *IEEE Transactions on Geoscience and Remote Sensing* **51** (2), 742–54 (2013).

20 M. Studinger, L. Koenig, S. Martin and J. Sonntag, 'Operation IceBridge: Using instrumented aircraft to bridge the observational gap between ICESat and ICESat-2', *IEEE International Geoscience and Remote Sensing Symposium* 1918–19 (2010).

21 L. Koenig, S. Martin, M. Studinger and J. Sonntag, 'Polar airborne observations fill gap in satellite data', *Eos Transactions of the American Geophysical Union* **91** (38), 333–4 (2010).

22 L. A. Plewes and B. Hubbard, 'A review of the use of radio-echo sounding in glaciology', *Progress in Physical Geography* **25** (2), 203–36 (2001); E. King, R. Hindmarsh, H. F. J. Corr and R. Bingham, presented at the International Symposium on Radioglaciology and its Applications, Madrid, Spain, 2008 (unpublished).

23 H. F. J. Corr, A. Jenkins, K. W. Nicholls and C. S. M. Doake, 'Precise measurement of changes in ice-shelf thickness by phase-sensitive radar to determine basal melt rates', *Geophysical Research Letters* **29** (8) (2002).

24 C. A. Cardenas Mansilla, M. Jenett, K. Schunemann and J. Winkelmann, 'Sub-ice topography in Patriot Hills, West Antarctica: first results of a newly developed high-resolution FM-CW radar system', *Journal of Glaciology* **56** (195), 162–6 (2010); J. A. Uribe, R. Zamora, G. Gacitua, A. Rivera and D. Ulloa, 'A low power consumption radar system for measuring ice thickness and snow/firn accumulation in Antarctica', *Annals of Glaciology* **55** (67), 39–48 (2014).

25 P. V. Brennan, L. B. Lok, K. Nicholls and H. Corr, 'Phase-sensitive FMCW radar system for high-precision Antarctic ice shelf profile monitoring', *IET Radar Sonar and Navigation* **8** (7), 776–86 (2014); K. W. Nicholls, H. F. J. Corr, C. L. Stewart, L. B. Lok, P. V. Brennan and D. G. Vaughan, 'A ground-based radar for measuring vertical strain rates and time-varying basal melt rates in ice sheets and shelves', *Journal of Glaciology* **61** (230), 1079–87 (2015).

26 L. B. Lok, M. Ash, K. W. Nicholls and P. V. Brennan, 'Autonomous phase-sensitive radio echo sounder for monitoring and imaging Antarctic ice shelves', 2015 8th International Workshop on Advanced Ground Penetrating Radar (IWAGPR), Firenze, Italy (2015).

27 D. J. Cavalieri, C. L. Parkinson, P. Gloersen and H. J. Zwally, 'Sea ice concentrations from Nimbus-7 SMMR and DMSP SSM/I-SSMIS passive microwave data [concentration]' (NASA DAAC at the National Snow and Ice Data Center, Boulder, Colorado, USA, 1996, updated yearly).

28 L. Brucker and T. Markus, 'Arctic-scale assessment of satellite passive microwave-derived snow depth on sea ice using Operation IceBridge airborne data', *Journal of Geophysical Research-Oceans* **118** (6), 2892–905 (2013); J. King, S. Howell, C. Derksen, N. Rutter, P. Toose, J. F. Beckers, C. Haas, N. Kurtz and J. Richter-Menge, 'Evaluation of Operation

IceBridge quick-look snow depth estimates on sea ice', *Geophysical Research Letters* **42** (21), 9302–10 (2015).
29 M. Morlighem, E. Rignot, J. Mouginot, X. Wu, H. Seroussi, E. Larour and J. Paden, 'High-resolution bed topography mapping of Russell Glacier, Greenland, inferred from Operation IceBridge data', *Journal of Glaciology* **59** (218), 1015–23 (2013).
30 N. T. Kurtz, S. L. Farrell, M. Studinger, N. Galin, J. P. Harbeck, R. Lindsay, V. D. Onana, B. Panzer and J. G. Sonntag, 'Sea ice thickness, freeboard, and snow depth products from Operation IceBridge airborne data', *The Cryosphere* **7** (4), 1035–56 (2013).
31 W. L. Qi and A. Braun, 'Accelerated elevation change of Greenland's Jakobshavn Glacier observed by ICESat and IceBridge', *IEEE Geoscience and Remote Sensing Letters* **10** (5), 1133–7 (2013).
32 E. M. Enderlin, I. M. Howat, S. Jeong, M.-J. Noh, J. H. van Angelen and M. R. van den Broeke, 'An improved mass budget for the Greenland ice sheet', *Geophysical Research Letters* **41** (3), 866–72 (2014).
33 T. J. Young, P. Christoffersen, K. W. Nicholls, L. B. Lok, S. H. Doyle, B. P. Hubbard, C. L. Stewart, C. Hofstede, M. Bougamont, J. A. Todd, P. V. Brennan and A. B. Hubbard, presented at the European Geosciences Union Meeting, Vienna, Austria, 2016 (unpublished).
34 R. C. Johnson, *Antenna Engineering Handbook* (McGraw-Hill, New York, 1993), 3rd edn.
35 Lok et al., 'Autonomous phase-sensitive radio echo sounder for monitoring and imaging Antarctic ice shelves'.
36 Nicholls et al., 'A ground-based radar for measuring vertical strain rates and time-varying basal melt rates in ice sheets and shelves'.
37 Lok et al., 'Autonomous phase-sensitive radio echo sounder for monitoring and imaging Antarctic ice shelves'.
38 J. P. Shonkoff and S. N. Bales, 'Science does not speak for itself: Translating child development research for the public and its policymakers', *Child Development* **82** (1), 17–32 (2011).
39 Tilling et al., 'Increased Arctic sea ice volume after anomalously low melting in 2013'.
40 S. F. Henley, A. M. Dolan, A. Pope, A. Kirchagessner and J. Gales, 'The UK Polar Network: Inspiring the next generation of Polar scientists in the UK and beyond', presented at the IPY 2012 Conference, Montreal, Canada, 2012 (unpublished).
41 K. Giles, 'Exploring the Arctic from space', UCL Lunch Hour Lecture Series. YouTube: https://www.youtube.com/watch?v=xYxyv8WUQjo
42 Giles et al., 'Western Arctic Ocean freshwater storage increased by wind-driven spin-up of the Beaufort Gyre', *Natural Geoscience* **5** (3), 194–7.

Chapter 5

1 Erica M. Dingman, 'Has Canada Shown its Arcticness?', *Connections: The Quarterly Journal* **10** (1), 24–45 (2010).
2 Charles K. Ebinger and Evie Zambetakis, 'The Geopolitics of Arctic Melt', *International Affairs* **85** (6), 1215–32 (November 2009); Valery Konyshev and Aleksandr Sergunin, 'The Arctic at the Crossroads of Geopolitical Interests', *Russian Politics and Law* **50** (2), 34–54, (1 March 2012); Iain Watson, 'Middle Power Alliances and the Arctic: Assessing Korea-UK Pragmatic Idealism', *Korea Observer* **45** (2), 275–320 (2014); Michał Łuszczuk, Piotr Graczyk, Adam Stępień and Małgorzata Śmieszek, 'Poland's Policy towards the Arctic: Key Areas and Priority Actions', *PISM Policy Paper*, no. 11 (113) (May 2015), https://www.pism.pl/files/?id_plik=19746; Ronald O'Rourke, *Changes in the Arctic; Background and Issues for Congress* (CRS Report No. R41153) (Washington, DC: Congressional Research Service, 2016), https://www.fas.org/sgp/crs/misc/R41153.pdf; Jingchao Peng and Njord Wegge, 'China's bilateral diplomacy in the Arctic', *Polar Geography* **38** (3), 233–49.
3 Andris Sprūds and Toms Rostoks (eds), *Perceptions and Strategies of Arcticness in Sub-Arctic Europe* (Riga: SIA Hansa Print Riga, 2014), 1.
4 Juha Ridanpää, 'A Masculinist Northern Wilderness and the Emancipatory Potential of Literary Irony', *Gender, Place & Culture* **17** (3), 319–35 (2010); Daniel Chartier, 'Representations of North and Winter. The methodological point of view of "nordicity" and "winterity"', in Enrique del Acebo Ibáñez and Helgi Gunnlaugsson (eds), *La circumpolaridad como fenómeno sociocultural. Pasado, presente, future* (Buenos Aires: Universidad de Buenos Aires, 2010), 36–7.

5 Ulrike Spring and Johan Schimanski, 'The Useless Arctic: Exploiting Nature in the Arctic in the 1870s', *Nordlit*, no. 35, 27–39 (2015).

6 Marthe T. Fjellestad, 'Picturing the Arctic', *Polar Geography* **39** (4), 228–38 (2016).

7 Philip E. Steinberg, Johanne Bruun and Ingrid A. Medby, 'Covering Kiruna: A Natural Experiment in Arctic Awareness', *Polar Geography* **37** (4), 273–97 (2014).

8 Leena S. Cho and Matthew G. Jull, 'Urbanized Arctic Landscapes: Critiques and Potentials from a Design Perspective', George Washington University Institute for European, Russian, and Eurasian Studies (2013), accessed 17 November 2016, https://www2.gwu.edu/~ieres-gwu/assets/docs/Cho&Jull_UrbanizedArcticLandscapes_final.pdf

9 Suzanne Robinson, '"Take it from the top": Northern conceptions about identity in the western Arctic and beyond', *Polar Record* **48** (3), 222–9, 223 (2012).

10 Robinson, 'Take it from the top', 224.

11 E. Carina H. Keskitalo et al., 'Contrasting Arctic and Mainstream Swedish Descriptions of Northern Sweden: The View from Established Domestic Research', *ARCTIC* **66** (3), 351–65, 353 (2013); Spring and Schimanski, 'The Useless Arctic', x. For a more comprehensive critique of this approach, see Lars Jensen, 'Greenland, Arctic Orientalism and the search for definitions of a contemporary postcolonial geography', *KULT – Postkolonial Temaserie* **12**, 139–53, 139–42 (2015); Ann Fienup-Riordan, *Freeze Frame: Alaska Eskimos in the Movies* (Seattle: University of Washington Press, 1995), xi–xiii.

12 Edward W. Said, *Orientalism*, 1st edn (New York: Vintage Books, 1979).

13 Ann Fienup-Riordan, *Freeze Frame: Alaska Eskimos in the Movies* (Seattle: University of Washington Press, 1995), xi–xiii.

14 E. Carina H. Keskitalo, *Negotiating the Arctic*, 1st edn (New York: Routledge, 2004).

15 Robert G. David, *The Arctic in the British Imagination, 1818–1914*, 1st edn (Manchester: Manchester University Press, 2000); Anka Ryall, Johan Schimanski and Henning Howlid Waerp, 'Arctic Discourses: An Introduction', in Johan Schimanski, Henning Howlid Waerp and Anka Ryall (eds), *Arctic Discourses* (Cambridge Scholars Publishing, 2010), ix–xxii..

16 Ryall, Schimanski and Waerp, 'Arctic Discourses', x.

17 Lisa Bloom, *Gender on Ice: American Ideologies of Polar Expeditions* (NED – New Edition., Vols. 1 10) (University of Minnesota Press, 1993), 57–109. Retrieved from http://www.jstor.org/stable/10.5749/j.ctttsm1t

18 Ridanpää, 'Masculinist Northern Wilderness', 319.

19 Ridanpää, 'Masculinist Northern Wilderness', 326; Bloom, *Gender on Ice*, 101–7; Spring and Schimanski, 'The Useless Arctic', 35–7.

20 Ridanpää, 'Masculinist Northern Wilderness', 326.

21 Hannes Gerhardt, Philip Steinberg, Jeremy Tasch, Sandra J. Fabiano and Rob Shields, 'Contested Sovereignty in a Changing Arctic', *Annals of the Association of American Geographers* **100** (4), 992–1002 (31 August 2010); Fjellestad, 'Picturing the Arctic', 237.

22 Norman F. Dixon, *On the Psychology of Military Incompetence*, 2016 edn (New York: Basic Books, 1976), 227.

23 Jean Comaroff and John Comaroff, *Ethnicity, Inc.* (Illinois: University of Chicago Press, 2009), 28. Retrieved from https://books.google.co.il/books?id=2efYCwDP6VsC&printsec=frontcover&source=gbs_ge_summary_r&cad=0#v=onepage&q&f=false

24 Comaroff and Comaroff, *Ethnicity, Inc.*, 9–10.

25 Kristín Loftsdóttir, 'The Exotic North: Gender, Nation Branding and Post-colonialism in Iceland', *NORA – Nordic Journal of Feminist and Gender Research* **23** (4), 246–60 (2015).

26 Ryall, Schimanski and Waerp, 'Arctic Discourses', xii.

27 Loftsdóttir, 'Exotic North', 255.

28 Loftsdóttir, 'Exotic North', 257.

29 Loftsdóttir, 'Exotic North', 253.

30 Annette Therkelsen and Henrik Halkier, 'Umbrella Place Branding. A Study of Friendly Exoticism and Exotic Friendliness in Coordinated National Tourism and Investment Promotion', *Discussion Papers: Center for International Studies* **26** (24), 1–24, 7 (2004).

31 Loftsdóttir, 'Exotic North', 252.

32 Lassi Heininen, 'State of the Arctic Strategies and Policies – A Summary', in Lassi Heininen, H. Exner-Pirot and J. Plouffe (eds), *Arctic Yearbook 2012* (Akureyri: Northern Research Forum, 2012), 2–47.

33 Monica Tennberg, 'Is Adaptation Governable in the Arctic? National and Regional Approaches to Arctic Adaptation Governance', in T. Koivurova, E. C. H. Keskitalo and

N. Bankes (eds), *Climate Governance in the Arctic*, Vol. 50 (Dordrecht: Springer Netherlands, 2009), 289–301, 289.

34 Golo M. Bartsch, 'Die Governance der Arktis: Akteure, Institutionen und politische Perspektiven im tauenden Hohen Norden' (The Governance of the Arctic: Actors, institutions and political perspectives in the melting high north) (Master's thesis, Fern-Universität in Hagen, 2011), 1–85. Retrieved from http://ecologic.eu/sites/files/cv/2013/bartsch_2011_ma_arktis_sicherheit.pdf; Tennberg, 'Adaptation Governable'.

35 Ingrid A. Medby, 'Arctic State, Arctic Nation? Arctic National Identity Among the Post-Cold War Generation in Norway', *Polar Geography* **37** (3), 252–69 (2014).

36 Frank Sejersen, *Rethinking Greenland and the Arctic in the Era of Climate Change: New Northern Horizons* (New York: Routledge, 2015), 7.

37 Nadine Fabbi, 'Inuit Political Engagement in the Arctic', in L. Heininen, H. Exner-Pirot and J. Plouffe (eds), *Arctic Yearbook 2012* (Akureyri: Northern Research Forum, 2012), 160–76, 163.

38 Fabbi, 'Inuit Political Engagement', 171.

39 Lincoln E. Flake, 'Forecasting Conflict in the Arctic: The Historical Context of Russia's Security Intentions', *The Journal of Slavic Military Studies* **28** (1), 72–98 (2015).

40 Ian G. Brosnan, Thomas M. Leschine and Edward L. Miles, 'Cooperation or Conflict in a Changing Arctic?', *Ocean Development & International Law* **42** (1–2), 173–210 (2011); Kristian Åtland, 'Interstate Relations in the Arctic: An Emerging Security Dilemma?', *Comparative Strategy* **33**(2), 145–66 (2014).

41 Michael Byers, 'Cold Peace: Arctic Cooperation and Canadian Foreign Policy', *International Journal: Canada's Journal of Global Policy Analysis* **65** (4), 899–912 (2010).

42 Piotr Kobza, 'Civilian Power Europe in the Arctic: How Far Can the European Union Go North?', EU Diplomacy Paper, College of Europe (2015), 1-30; P. Whitney Lackenbauer, 'Canada and the Asian Observers to the Arctic Council: Anxiety and Opportunity', *Asia Policy* **18** (1), 22–9 (2014); Njord Wegge, 'The Political Order in the Arctic: Power Structures, Regimes and Influence', *Polar Record* **47** (2), 165–76 (2011).

43 Margaret Blunden, 'Geopolitics and the Northern Sea Route', *International Affairs* **88** (1), 115–29 (2012).

44 Taehwan Kim, 'Paradigm Shift in Diplomacy: A Conceptual Model for Korea's "New Public Diplomacy"', *Korea Observer* **43** (4), 527–55 (2012); Watson, 'Middle Power Alliances', 280.

45 Watson, 'Middle Power Alliances', 277.

46 Sandra Maria Rodrigues Balão, 'The European Union's Arctic Strategy(ies): The Good and/or the Evil?', in Lassi Heininen (ed.), *Security and Sovereignty in the North Atlantic* (Basingstoke: Palgrave Macmillan, 2014).

47 E. Carina H. Keskitalo, *Negotiating the Arctic*.

48 Sejersen, *Rethinking Greenland*, 5–11.

49 A term coined by Benedict Anderson, *Imagined Communities: Reflections on the Origin and Spread of Nationalism* (London: Verso, 1983).

50 Toms Rostoks. 'Conclusion: What role is there for sub-Arctic states in the Arctic's Future?', in Andris Sprūds and Toms Rostoks (eds), *Perceptions and Strategies of Arcticness in Sub-Arctic Europe* (Riga: Latvian Institute of International Affairs, 2014), 227.

51 Rostoks, 'Conclusion', 218.

52 Alexander Wendt, 'Anarchy is What States Make of it: The Social Construction of Power Politics', *International Organization* **46** (2), 391–425 (1992).

53 Michel Foucault, *Archaeology of Knowledge*, 1st edn (London: Routledge, 1972).

54 Spring and Schimanski, 'The Useless Arctic', 14.

Chapter 7

1 Qinghai-Tibetan Plateau, China and Tibet are used interchangeably in this chapter to refer to areas with pastoralists of Tibetan ethnicity.

2 Marius W. Næss, 'Climate Change, Risk Management and the End of *Nomadic* Pastoralism', *International Journal of Sustainable Development and World Ecology* **20** (2), 123–33 (2013).

3 Emily T. Yeh, 'Tibetan Range Wars: Spatial Politics and Authority on the Grasslands of Amdo', *Development and Change* **34** (3), 499–523 (2003).

4 Miaogen Shen, Shilong Piao, Su-Jong Jeong, Liming Zhou, Zhenzhong Zeng, Philippe Ciais, Deliang Chen, et al., 'Evaporative Cooling over the Tibetan Plateau Induced by Vegetation Growth', *Proceedings of the National Academy of Sciences* **112** (30), 9299–304 (28 July 2015).
5 H. I. Heikkinen, S. Sarkki and M. Nuttall, 'Users or Producers of Ecosystem Services? A Scenario Exercise for Integrating Conservation and Reindeer Herding in Northeast Finland', *Pastoralism: Research, Policy and Practice* **2** (11) (2012).
6 J. G. Galaty and P. C. Salzman, *Change and Development in Nomadic and Pastoral Societies*. International Studies in Sociology and Social Anthropology Vol. 33 (Leiden: Brill, 1981); Caroline Humphrey and David Sneath, *The End of Nomadism? Society, State, and the Environment in Inner Asia* (Durham, NC: Duke University Press, 1999).
7 E. Mwangi and E. Ostrom, 'A Century of Institutions and Ecology in East Africa's Rangelands: Linking Institutional Robustness with the Ecological Resilience of Kenya's Maasailand', in V. Beckmann and M. Padmanabhan (eds), *Institutions and Sustainability – Political Economy of Agriculture and the Environment – Essays in Honour of Konrad Hagedorn* (Netherlands: Springer, 2009), 195–221.
8 Garrett J. Hardin, 'The Tragedy of the Commons', *Science* **162**, 1243–48 (1968).
9 Mwangi and Ostrom, 'A Century of Institutions and Ecology in East Africa's Rangelands'.
10 Jean Ensminger and Andrew Rutten, 'The Political Economy of Changing Property Rights: Dismantling a Pastoral Commons', *American Ethnologist* **18** (4), 683–99 (1991).
11 H. Lamprey, 'Pastoralism Yesterday and Today: The Overgrazing Problem', in F. Bourliers (ed.), *Ecosystems of the World 13: Tropical Savannas* (Amsterdam: Elsevier Scientific Publishing Co., 1983).
12 P. Ho, 'The Clash over State and Collective Property: The Making of the Rangeland Law', *The China Quarterly* **161**, 240–63 (2000).
13 R. B. Harris, 'Rangeland Degradation on the Qinghai-Tibetan Plateau: A Review of the Evidence of Its Magnitude and Causes', *Journal of Arid Environments* **74** (1), 1–12 (2010).
14 Robin S. Reid, María E. Fernández-Giménez and Kathleen A. Galvin, 'Dynamics and Resilience of Rangelands and Pastoral Peoples around the Globe', *Annual Review of Environment and Resources* **39** (1), 217–42 (2014).
15 Anonymous, 'St.Prp. Nr. 63: Om Reindriftsavtalen 2008/2009 Og Om Endringer I Statsbudsjettet for 2008 M.M', 24: Det Kongelige Landbruks- og Matdepartement, 2008; Anonymous, 'Ressursregnskap for Reindriftsnæringen' (Alta, Norway: Reindriftsforvaltningen, 2008), 164.
16 Geoff Kushnick, Russell D. Gray and Fiona M. Jordan, 'The Sequential Evolution of Land Tenure Norms', *Evolution and Human Behavior* **35** (4), 309–18 (2014).
17 F. Lamptey, 'Participatory GIS Tools for Mapping Indigenous Knowledge in Customary Land Tenure Dynamics: Case of Peri-Urban Northern Ghana', Master of Science, International Institute for Geo-information Science and Earth Observation, Enschede, Netherlands, 2009.
18 E. A. Smith, 'Risk and Uncertainty in the "Original Affluent Society": Evolutionary Ecology of Resource Sharing and Land Tenure', in T. Ingold, D. Riches and J. Woodburn (eds), *Hunters and Gatherers: History, Evolution, and Social Change* (Oxford: Berg, 1988), 222–52.
19 N. E. Levine, 'Cattle and the Cash Economy: Responses to Change among Tibetan Nomadic Pastoralists in Sichuan, China', *Human Organization* **58** (2), 161–72 (1999); Levine, 'From Nomads to Ranchers: Managing Pasture among Ethnic Tibetans in Sichuan', Paper presented at the Development, Society and Environment in Tibet, Vienna, 1995; B. Gelek, 'The Washu Serthar: A Nomadic Community of Eastern Tibet', in G. E. Clarke (ed.), *Development, Society, and Environment in Tibet* (Vienna: Verlag der Österreichischen Akademie der Wissenschaften, 1998), 47–58; Marius W. Næss, 'Living with Risk and Uncertainty: The Case of the Nomadic Pastoralists in the Aru Basin, Tibet', Candidatus Rerum Politicarum, University of Tromsø, 2003.
20 Ragnar Nilsen and Jens Halvdan Mosli, *Inn Fra Vidda: Hushold Og Økonomisk Tilpasning I Reindrifta I Guovdageaidnu 1960–1993*. Norut Samfunnsforskning Rapport. Guovdageaidnu: Bajos (in Norwegian) (1994); Robert Paine, *Herds of the Tundra: A Portrait of Saami Reindeer Pastoralism*. Smithsonian Series in Ethnographic Inquiry (Washington, DC & London: Smithsonian Institution Press, 1994); Robert N. Pehrson, *The Bilateral Network of Social Relations in Könkämä Lapp District*. Samiske Samlinger 7 (Oslo: Universitetsforlaget, 1964).
21 Yeh, 'Tibetan Range Wars: Spatial Politics and Authority on the Grasslands of Amdo'.

22 Levine, 'From Nomads to Ranchers: Managing Pasture among Ethnic Tibetans in Sichuan'; Yeh, 'Tibetan Range Wars: Spatial Politics and Authority on the Grasslands of Amdo'; F. Pirie, 'Segmentation within the State: The Reconfiguration of Tibetan Tribes in China's Reform Period', *Nomadic Peoples* 9 (1–2) (2005).

23 Levine, 'From Nomads to Ranchers: Managing Pasture among Ethnic Tibetans in Sichuan'.

24 Næss, 'Living with Risk and Uncertainty: The Case of the Nomadic Pastoralists in the Aru Basin, Tibet'; M. C. Goldstein and C. M. Beall, *Nomads of Western Tibet: The Survival of a Way of Life* (London: Serindia Publications, 1990).

25 Næss, 'Living with Risk and Uncertainty: The Case of the Nomadic Pastoralists in the Aru Basin, Tibet'.

26 Næss, 'Climate Change, Risk Management and the End of *Nomadic* Pastoralism'.

27 A. F. Marin, 'Confined and Sustainable? A Critique of Recent Pastoral Policy for Reindeer Herding in Finnmark, Northern Norway', *Nomadic Peoples* **10** (2), 209–32 (2006).

28 Paine, *Herds of the Tundra: A Portrait of Saami Reindeer Pastoralism*.

29 J. Å. Riseth, 'Sámi Reindeer Management under Technological Change 1960–1990: Implications for Common-Pool Resource Use under Various Natural and Institutional Conditions: A Comparative Analysis of Regional Development Paths in West Finnmark, North Trøndelag, and South Trøndelag/Hedmark, Norway', Doctor scientarium, Norges landbrukshøgskole, 2000; Paine, *Herds of the Tundra: A Portrait of Saami Reindeer Pastoralism*; Andrei Marin and Ivar Bjørklund, *A Tragedy of Errors? Institutional Dynamics and Land Tenure in Finnmark, Norway*. International Journal of the Commons **9** (1), 19–40. (2015).

30 Goldstein and Beall, *Nomads of Western Tibet: The Survival of a Way of Life*; Næss, 'Living with Risk and Uncertainty: The Case of the Nomadic Pastoralists in the Aru Basin, Tibet'.

31 Næss, 'Living with Risk and Uncertainty: The Case of the Nomadic Pastoralists in the Aru Basin, Tibet'.

32 Næss, 'Living with Risk and Uncertainty: The Case of the Nomadic Pastoralists in the Aru Basin, Tibet'.

33 J. Å Riseth and A. Vatn, 'Modernization and Pasture Degradation: A Comparative Study of Two Sàmi Reindeer Pasture Regions in Norway', *Land Economics* **85** (1), 87–106 (2009).

34 Marin, 'Confined and Sustainable? A Critique of Recent Pastoral Policy for Reindeer Herding in Finnmark, Northern Norway'.

35 Marin, 'Confined and Sustainable? A Critique of Recent Pastoral Policy for Reindeer Herding in Finnmark, Northern Norway'.

36 Næss, 'Climate Change, Risk Management and the End of *Nomadic* Pastoralism'.

37 Næss, 'Climate Change, Risk Management and the End of *Nomadic* Pastoralism'.

38 T. Banks, 'Property Rights and the Environment in Pastoral China: Evidence from the Field', *Development and Change* **32** (4), 717–40 (2001).

39 D. M. Williams, 'Grassland Enclosures: Catalyst of Land Degradation in Inner Mongolia', *Human Organization* **55** (3), 307–13 (1996).

40 Z. Yan, N. Wu, Y. Dorji and R. Jia, 'A Review of Rangeland Privatisation and Its Implications in the Tibetan Plateau, China', *Nomadic Peoples* **9** (1&2), 31–51 (2005).

41 Næss, 'Climate Change, Risk Management and the End of *Nomadic* Pastoralism'.

42 N. E. Levine, 'Reconstructing Tradition: Persistence and Change in Golog Social Structure', unpublished result.

43 Næss, 'Living with Risk and Uncertainty: The Case of the Nomadic Pastoralists in the Aru Basin, Tibet'; Næss, 'Climate Change, Risk Management and the End of *Nomadic* Pastoralism'.

44 Anonymous, 'Lov Om Reindrift Av 15. Juni. 2007 Nr 40', 2007.

45 Anonymous, 'Ressursregnskap for Reindriftsnæringen', p.156. (Alta, Norway: Reindriftsforvaltningen, 2006); Anonymous, 'Årsrapport 2007 (Yearly Report 2007)', p. 31 (in Norwegian) (Alta, Norway: Reindriftsforvaltningen, 2007).

46 A. Berg, 'På Riktig Spor', *Reindriftsnytt* **41** (1), 6–7 (2007); Anonymous, 'Lov Om Reindrift-Av 15. Juni. 2007 Nr 40'.

47 Reid et al., 'Dynamics and Resilience of Rangelands and Pastoral Peoples around the Globe'.

48 Encroachment – due to activities like mining, oil and gas extraction, forestry and wind- and hydro-electric power production – is steadily increasing in northern ecosystems.

49 K. A. Galvin, 'Transitions: Pastoralists Living with Change', *Annual Review of Anthropology* **38** (1), 185–98 (2009).

50 Næss, 'Climate Change, Risk Management and the End of *Nomadic* Pastoralism'.

51 K. Bauer, 'Development and the Enclosure Movement in Pastoral Tibet since the 1980s', *Nomadic Peoples* **9** (1–2), 53–81 (2005); in Yushu Tibetan autonomous prefecture (Qinghai Province) the area of fenced in rangelands changed from 33,370 ha in 1983 to 966,000 ha in 2005. Keep in mind that while it indicates a substantial increase, it represents only 16.9% of Yushu's utilisable rangeland; A. Gruschke, 'Tibetan Pastoralists in Transition. Political Change and State Interventions in Nomad Societies', in H. Kreutzmann (ed.), *Pastoral Practices in High Asia, Agency of 'Development' Effected by Modernisation, Resettlement and Transformation*, Advances in Asian Human-Environmental Research (Netherlands: Springer, 2012), 273–89.

52 Reid et al., 'Dynamics and Resilience of Rangelands and Pastoral Peoples around the Globe'.

53 Williams, 'Grassland Enclosures: Catalyst of Land Degradation in Inner Mongolia'.

54 R. H. Behnke, I. Scoones and C. Kerven (eds), *Range Ecology at Disequilibrium. New Models of Natural Variability and Pastoral Adaptation in African Savannas* (London: Overseas Development Institute, 1993).

55 J. M. Milner, D. A. Elston and S. D. Albon, 'Estimating the Contributions of Population Density and Climatic Fluctuations to Interannual Variation in Survival of Soay Sheep', *Journal of Animal Ecology* **68** (6), 1235–47 (1999).

56 B.-J. Bårdsen and T. Tveraa, 'Density Dependence Vs. Density Independence – Linking Reproductive Allocation to Population Abundance and Vegetation Greenness', *Journal of Animal Ecology* **81** (2), 364–76 (2012).

57 Carol Kerven, Bernd Steimann, Chad Dear and Laurie Ashley, 'Researching the Future of Pastoralism in Central Asia's Mountains: Examining Development Orthodoxies', *Mountain Research and Development* **32** (3), 368–77 (2012).

58 Bård-Jørgen Bårdsen, Marius Warg Næss, T. Tveraa, P. Fauchald and K. Langeland, 'Risk-Sensitive Reproductive Allocation: Fitness Consequences of Body Mass Losses in Two Contrasting Environments', *Ecology and Evolution* **4** (7), 1030–8 (2014).

59 E. Aslaksen and N. H. Måsø, 'Venter På Døden', *NRK Sápmi – NRK*, 2010.

60 Aslaksen and Måsø, 'Venter På Døden'.

61 V. H. Hausner, P. Fauchald, T. Tveraa, E. Pedersen, J.-L. L. Jernsletten, B. Ulvevadet, R. A. Ims, N. Yoccoz and K. A. Bråthen, 'The Ghost of Development Past: The Impact of Economic Security Policies on Saami Pastoral Ecosystems', *Ecology and Society* **16** (3), 4 (2011).

62 Reid et al., 'Dynamics and Resilience of Rangelands and Pastoral Peoples around the Globe'.

63 E. E. Evans-Pritchard, *The Nuer: a Description of the Modes of Livelihood and Political Institutions of a Nilotic People* (Oxford: Clarendon Press, 1940); D. J. Stenning, 'Transhumance, Migratory Drift, Migration; Patterns of Pastoral Fulani Nomadism', *Journal of the Royal Anthropological Institute of Great Britain and Ireland* **87**, 57–75 (1957); Stenning, *Savannah Nomads: A Study of the Wodaabe Pastoral Fulani of Western Bornu Province Northern Region, Nigeria* (London: Published for the International African Institute by Oxford University Press, 1959); Neville Dyson-Hudson, *Karimojong Politics* (Oxford: Clarendon Press, 1966); Dyson-Hudson, 'The Study of Nomads', in W. Irons and N. Dyson-Hudson (eds), *Perspectives on Nomadism* (Leiden, Netherlands: E. J. Brill, 1972), 2–29.

64 Riseth and Vatn, 'Modernization and Pasture Degradation: A Comparative Study of Two Sàmi Reindeer Pasture Regions in Norway'.

65 R. Pape and J. Loffler, 'Climate Change, Land Use Conflicts, Predation and Ecological Degradation as Challenges for Reindeer Husbandry in Northern Europe: What Do We Really Know after Half a Century of Research?', *AMBIO: A Journal of the Human Environment* **41**, 421–34 (2012).

66 G. Oba and W. J. Lusigi, 'An Overview of Drought Strategies and Land Use in African Pastoral Systems', *Network Paper* (London: Overseas Development Institute, 1987).

67 Oba and Lusigi, 'An Overview of Drought Strategies and Land Use in African Pastoral Systems'.

68 Næss, 'Climate Change, Risk Management and the End of *Nomadic* Pastoralism'.

69 A. Agrawal, 'Mobility and Cooperation among Nomadic Shepherds – the Case of the Raikas', *Human Ecology* **21** (3), 261–79 (1993).

70 M. E. Fernandez-Gimenez and S. Le Febre, 'Mobility in Pastoral Systems: Dynamic Flux or Downward Trend?', *International Journal of Sustainable Development and World Ecology* **13** (5), 341–62 (2006).

71 P. D. Little, K. Smith, B. A. Cellarius, D. L. Coppock and C. B. Barrett, 'Avoiding Disaster: Diversification and Risk Management among East African Herders', *Development and Change* **32** (3), 401–33 (2001).

72 Fernandez-Gimenez and Le Febre, 'Mobility in Pastoral Systems: Dynamic Flux or Downward Trend?'
73 Humphrey and Sneath, *The End of Nomadism? Society, State, and the Environment in Inner Asia.*
74 H. J. Schwartz, 'Ecological and Economic Consequences of Reduced Mobility in Pastoral Livestock Production Systems', in E. Fratkin and E. A. Roth (eds), *As Pastoralists Settle: Social, Health, and Economic Consequences of Pastoral Sedentarization in Marsabit District, Kenya*. Studies in Human Ecology and Adaptation (New York and London: Kluwer Academic Publishers, 2005), 69–86.
75 D. Ojima and T. Chuluun, 'Policy Changes in Mongolia: Implications for Land Use and Landscapes', in K. A. Galvin, R. S. Reid, R. H. Behnke, Jr and N. T. Hobbs (eds), *Fragmentation in Semi-Arid and Arid Landscapes: Consequences for Human and Natural Systems* (Dordrecht: Springer, 2008), 179–93; D. Sneath, 'Land Use, the Environment and Development in Post-Socialist Mongolia', *Oxford Development* **31** (4), 441–59 (2003); Williams, 'Grassland Enclosures: Catalyst of Land Degradation in Inner Mongolia'.
76 J. J. Cao, Y. C. Xiong, J. Sun, W. F. Xiong and G. Z. Du, 'Differential Benefits of Multi- and Single-Household Grassland Management Patterns in the Qinghai-Tibetan Plateau of China', *Human Ecology* **39** (2), 217–27 (2011).
77 J. J. Cao, Emily T. Yeh, N. M. Holden, Y. Yang and G. Du, 'The Effects of Enclosures and Land-Use Contracts on Rangeland Degradation on the Qinghai-Tibetan Plateau', *Journal of Arid Environments* **97**, 3–8 (2013).
78 W. J. Li, S. H. Ali and Q. Zhang, 'Property Rights and Grassland Degradation: A Study of the Xilingol Pasture, Inner Mongolia, China', *Journal of Environmental Management* **85** (2), 461–70 (2007).
79 J. L. Taylor, 'Negotiating the Grassland: The Policy of Pasture Enclosures and Contested Resource Use in Inner Mongolia', *Human Organization* **65** (4), 374–86 (2006).
80 John D. Farrington, 'De-Development in Eastern Kyrgyzstan and Persistence of Semi-Nomadic Livestock Herding', *Nomadic Peoples* **9** (1/2), 171–97 (2005).
81 D. Sneath, 'Ecology – State Policy and Pasture Degradation in Inner Asia', *Science* **281** (5380), 1147–8 (1998).
82 Gelek, 'The Washu Serthar: A Nomadic Community of Eastern Tibet.'; Levine, 'Cattle and the Cash Economy: Responses to Change among Tibetan Nomadic Pastoralists in Sichuan, China'; Levine, 'From Nomads to Ranchers: Managing Pasture among Ethnic Tibetans in Sichuan'; Levine, 'Reconstructing Tradition: Persistence and Change in Golog Social Structure'; Yeh, 'Tibetan Range Wars: Spatial Politics and Authority on the Grasslands of Amdo'; Pirie, 'Segmentation within the State: The Reconfiguration of Tibetan Tribes in China's Reform Period'; Paine, *Herds of the Tundra: A Portrait of Saami Reindeer Pastoralism*.
83 Michael S. Alvard, 'Kinship, Lineage, and an Evolutionary Perspective on Cooperative Hunting Groups in Indonesia', *Human Nature* **14** (2), 129–63 (2003); Ashleigh S. Griffin and Stuart A. West, 'Kin Selection: Fact and Fiction', *Trends in Ecology & Evolution* **17** (1), 15–21 (2002); R. L. Trivers, 'Evolution of Reciprocal Altruism', *Quarterly Review of Biology* **46** (1), 35–7 (1971); Eric A. Smith, 'Human Cooperation: Perspectives from Behavioral Ecology', in Peter Hammerstein (ed.), *Genetic and Cultural Evolution of Cooperation* (Cambridge, MA: MIT Press, 2003), 401–27; H. Gintis, S. Bowles, R. Boyd and E. Fehr, *Moral Sentiments and Material Interests: The Foundations of Cooperation in Economic Life*. Economic Learning and Social Evolution (Cambridge, MA: MIT Press, 2005); Marius W. Næss, B.-J. Bårdsen and T. Tveraa, 'Wealth-Dependent and Interdependent Strategies in the Saami Reindeer Husbandry, Norway', *Evolution and Human Behavior* **33** (6), 696–707 (2012).
84 Marius W. Næss, 'Cooperative Pastoral Production: Reconceptualizing the Relationship between Pastoral Labor and Production', *American Anthropologist* **114** (2), 309–21 (2012).
85 Robert Paine, *Camps of the Tundra: Politics through Reindeer among Saami Pastoralists*. Instituttet for Sammenlignende Kulturforskning. Serie B, Skrifter (Oslo: Instituttet for Sammenlignende Kulturforskning, 2009).
86 Levine, 'Reconstructing Tradition: Persistence and Change in Golog Social Structure'.
87 Næss, 'Cooperative Pastoral Production: Reconceptualizing the Relationship between Pastoral Labor and Production'.
88 Marius W. Næss, Bård-Jørgen Bårdsen, Per Fauchald and Torkild Tveraa, 'Cooperative Pastoral Production – the Importance of Kinship', *Evolution and Human Behavior* **31** (4), 246–58 (2010); Marius W. Næss, Per Fauchald and Torkild Tveraa. 'Scale Dependency and the "Marginal" Value of Labor', *Human Ecology* **37** (2), 193–211 (Apr 2009).

89 T. Yamaguchi, 'Transition of Mountain Pastoralism: An Agrodiversity Analysis of the Livestock Population and Herding Strategies in Southeast Tibet, China', *Human Ecology* **39**, 141–54 (2011).

90 C. Athena Aktipis, 'Is Cooperation Viable in Mobile Organisms? Simple Walk Away Rule Favors the Evolution of Cooperation in Groups', *Evolution And Human Behavior* **32** (4), 263–76 (2011).

91 Charles Efferson, Carlos P. Roca, Sonja Vogt and Dirk Helbing, 'Sustained Cooperation by Running away from Bad Behavior', *Evolution and Human Behavior* **37** (1), 1–9 (2016).

92 P. Ho, 'China's Rangelands under Stress: A Comparative Study of Pasture Commons in the Ningxia Hui Autonomous Region', *Development and Change* **31** (2), 385–412 (2000); Cao et al., 'The Effects of Enclosures and Land-Use Contracts on Rangeland Degradation on the Qinghai-Tibetan Plateau'; Pirie, 'Segmentation within the State: The Reconfiguration of Tibetan Tribes in China's Reform Period'.

93 Yeh, 'Tibetan Range Wars: Spatial Politics and Authority on the Grasslands of Amdo'.

94 Taylor, 'Negotiating the Grassland: The Policy of Pasture Enclosures and Contested Resource Use in Inner Mongolia'.

95 W. J. Li and L. Huntsinger, 'China's Grassland Contract Policy and Its Impacts on Herder Ability to Benefit in Inner Mongolia: Tragic Feedbacks', *Ecology and Society* **16** (2) (2011).

96 Taylor, 'Negotiating the Grassland: The Policy of Pasture Enclosures and Contested Resource Use in Inner Mongolia'.

97 D. M. Williams, 'Grassland Enclosures: Catalyst of Land Degradation in Inner Mongolia', *Human Organization* **55** (3), 307–13 (1996).

98 Arjun Appadurai, 'Theory in Anthropology: Center and Periphery', *Comparative Studies in Society and History* **28** (2), 356–74 (1986).

99 Appadurai, 'Theory in Anthropology: Center and Periphery'.

100 Appadurai, 'Theory in Anthropology: Center and Periphery'.

101 Henrik Erdman Vigh and David Brehm Sausdal, 'From Essence Back to Existence: Anthropology Beyond the Ontological Turn', *Anthropological Theory* **14** (1), 49–73 (2014).

102 Vigh and Sausdal, 'From Essence Back to Existence: Anthropology Beyond the Ontological Turn'.

103 D. Coumou and S. Rahmstorf, 'A Decade of Weather Extremes', *Nature Climate Change* **2** (7), 491–6 (2012).

104 I. Brannlund and P. Axelsson, 'Reindeer Management During the Colonization of Sami Lands: A Long-Term Perspective of Vulnerability and Adaptation Strategies', *Global Environmental Change: Human and Policy Dimensions* **21** (3), 1095–105 (2011).

105 M. Nori, M. Taylor and A. Sensi, 'Browsing on Fences: Pastoral Land Rights, Livelihoods and Adaptation to Climate Change', *Issue paper*, 29 (Nottingham, UK: International Institute for Environment and Development, 2008).

106 R. Hatfield and J. Davies, 'Global Review of the Economics of Pastoralism', *The World Initiative for Sustainable Pastoralism*, 44 (Nairobi: IUCN, 2006).

107 P. J. Blackwell, 'East Africa's Pastoralist Emergency: Is Climate Change the Straw that Breaks the Camel's Back?', *Third World Quarterly* **31** (8), 1321–38 (2010).

108 Nori et al., 'Browsing on Fences: Pastoral Land Rights, Livelihoods and Adaptation to Climate Change'.

109 Marius W. Næss and Bård-Jørgen Bårdsen, 'Environmental Stochasticity and Long-Term Livestock Viability: Herd-Accumulation as a Risk Reducing Strategy', *Human Ecology* **38** (1), 3–17 (2010); Næss and Bårdsen, 'Why Herd Size Matters – Mitigating the Effects of Livestock Crashes', *PLoS One* **8** (8), e70161 (2013).

110 Næss and Bårdsen, 'Environmental Stochasticity and Long-Term Livestock Viability: Herd-Accumulation as a Risk Reducing Strategy'.

111 Næss and Bårdsen, 'Why Herd Size Matters – Mitigating the Effects of Livestock Crashes'.

112 Michael Bollig and Barbara Göbel, 'Risk, Uncertainty and Pastoralism: An Introduction', *Nomadic Peoples* **1** (1), 5–21 (1997).

113 Næss, 'Cooperative Pastoral Production: Reconceptualizing the Relationship between Pastoral Labor and Production'; Næss et al., 'Cooperative Pastoral Production – the Importance of Kinship'; Næss et al., 'Wealth-Dependent and Interdependent Strategies in the Saami Reindeer Husbandry, Norway'; Næss et al., 'Scale Dependency and the "Marginal" Value of Labor'.

Chapter 8

1 D. L. Gautier, K. J. Bird, R. R. Charpentier, A. Grantz, D. W. Houseknecht, T. R. Klett, et al., 'Assessment of oil and undiscovered gas in the Arctic', *Science* **324**, 1175–79 (2009).
2 J. Vidal, '"Extraordinarily hot" Arctic temperatures alarm scientists', *The Guardian* (22 November 2016). Available at: https://www.theguardian.com/environment/2016/nov/22/extraordinarily-hot-arctic-temperatures-alarm-scientists (last accessed October 2016).
3 Arctic Council, *Arctic Resilience Report* (2016). Available at: http://arctic-council.org/arr/resources/project-publications/ (last accessed November 2016).
4 Nicole C. Kibert, 'Green Justice: A Holistic Approach to Environmental Injustice', *Journal of Land Use and Environmental Law* **17**(1), 169 (2001).
5 Julian Agyeman, Peter Cole, Randolph Haluza-DeLay and Pat O'Riley, *Speaking for Ourselves: Environmental Justice in Canada* (Vancouver, BC: UBC Press, 2010).
6 See Ryan Holifield, 'Environmental Justice as Recognition and Participation in Risk Assessment: Negotiating and Translating Health Risk at a Superfund Site in Indian Country', *Annals of the Association of American Geographers* **102** (3), 591–613 (2012); see also David Schlosberg and David Carruthers, 'Indigenous Struggles, Environmental Justice, and Community Capabilities', *Global Environmental Politics* **10** (4), 12–35 (2010); Liere Urkidi and Mariana Walter, 'Dimensions of Environmental Justice in Anti-gold Mining Movements in Latin America', *Geoforum* **42**, 683–95 (2011).
7 Chun-Chieh Chi, 'Capitalist Expansion and Indigenous Land Rights: Emerging Environmental Justice Issues in Taiwan', *The Asia Pacific Journal of Anthropology* **2** (2), 135–53 (2001).
8 See David A. McDonald, *Environmental Justice in South Africa* (Ohio: University Press, 2002); see also Gustav Etienne Visser, 'Spatialities of social justice and local government transition: Notes on and for a South African social justice discourse', *South African Geographical Journal* **85** (2), 99–111 (2003).
9 David Schlosberg, 'Theorising Environmental Justice: The Expanding Sphere of a Discourse', *Environmental Politics* **22** (1), 37 (2013).
10 Laura Pulido, 'Rethinking Environmental Racism: White Privilege and Urban Development in Southern California', *Annals of the Association of American Geographers* **90** (1), 12–40 (2000).
11 Susan Buckingham and Rakibe Kulcur, 'Gendered Geographies of Environmental Injustice', *Antipode* **41** (4), 659–83 (2009).
12 Mei-Fang Fan, 'Environmental Justice and Nuclear Waste Conflicts in Taiwan', *Environmental Politics* **15** (3), 417–34 (2006).
13 Charles R. Beitz, 'Rawls's Law of Peoples', *Ethics* **110** (4), 669–96 (2000); Thomas Nagel, 'The Problem of Global Justice', *Philosophy & Public Affairs* **33** (2), 113–47 (2005).
14 See Alex Aylett, 'Conflict, Collaboration and Climate Change: Participatory Democracy and Urban Environmental Struggles in Durban, South Africa', *International Journal of Urban and Regional Research* **34** (3), 478–95 (2010); see also Stephen M. Gardiner, 'Ethics and Global Climate Change', *Ethics* **114** (3), 555–600 (2004).
15 Nancy Fraser, *Justice Interruptus* (London: Routledge, 2014).
16 Harriet Bulkeley, Joann Carmin, Vanesa Castán Broto, Gareth A. S. Edwards and Sara Fuller, 'Climate Justice and Global Cities: Mapping the Emerging Discourses', *Global Environmental Change* **23** (5), 914–25 (2013).
17 R. J. Heffron, D. McCauley and B. K. Sovacool, 'Resolving Society's Energy Trilemma through the Energy Justice Metric', *Energy Policy* **87**, 168–176 (2015).
18 David Schlosberg, 'The Justice of Environmental Justice: Reconciling Equity, Recognition, and Participation in a Political Movement', in Andrew Light and Avner De-Shalit (eds), *Moral and Political Reasoning in Environmental Practice* (London: MIT Press, 2003), 125–56.
19 Nancy Fraser, 'Social justice in the age of identity politics', in George Henderson (ed.), *Geographical Thought: A Praxis Perspective* (London: Routledge, 1999), 56–89; Schlosberg, 'The Justice of Environmental Justice'.
20 See Gordon Walker, 'Beyond Distribution and Proximity: Exploring the Multiple Spatialities of Environmental Justice', *Antipode* **41** (4), 614–36 (2009); see also Robert D. Bullard, 'Environmental Justice in the 21st Century', in John S. Dryzek and David Schlosberg (eds), *Debating the Earth* (Oxford: Oxford University Press, 2005), 332–56.
21 Anna R. Davies, 'Environmental Justice as Subtext Or Omission: Examining Discourses of Anti-Incineration Campaigning in Ireland', *Geoforum* **37** (5), 708–24 (2006).

22 Helen Todd and Christos Zografos, 'Justice for the Environment: Developing an Indicator of Environmental Justice for Scotland', *Environmental Values* **43** (4), 483–501 (2005).
23 Maja Due Kadenic, 'Socioeconomic Value Creation and the Role of Local Participation in Large-Scale Mining Projects in the Arctic', *The Extractive Industries and Society* **2** (3), 562–71 (2015).
24 Brenda L. Parlee, 'Avoiding the Resource Curse: Indigenous Communities and Canada's Oil Sands', *World Development* **74**, 425–36 (2015).
25 Emma J. Stewart, J. Dawson and Dianne Draper. 'Cruise Tourism and Residents in Arctic Canada: Development of a Resident Attitude Typology', *Journal of Hospitality and Tourism* **18** (1), 95–106 (2011).
26 Darren A. McCauley, Raphael J. Heffron, Hannes Stephan and Kirsten Jenkins, 'Advancing Energy Justice: The Triumvirate of Tenets', *International Energy Law Review* **32** (3), 107–10 (2013).
27 R. Scott Marshall and Darrell Brown, 'Corporate Environmental Reporting: What's in a Metric?', *Business Strategy and the Environment* **12** (2), 87 (2003).
28 R. Edward Freeman, 'Stakeholder Management: Framework and Philosophy', in Robert A. Phillips and R. Edward Freeman (eds), *Stakeholders* (Cheltenham: Edward Elgar Publishing Limited, 1984), 1–33.
29 Mark Starik, 'Should trees have managerial standing? Towards stakeholder status for non-human nature', *Journal of Business Ethics* **14** (3), 207–17 (1995).
30 Jürgen Gerhard and Dieter Rucht, 'Mesomobilisation: Organizing and Framing in Two Protest Campaigns in West Germany', *American Journal of Sociology* **98** (3), 555–96 (1992).
31 David A. Snow, 'Framing Processes, Ideology and Discursive Fields', in David Snow, Sarah A. Soule and Hanspeter Kriesi (eds), *The Blackwell Companion to Social Movements* (Oxford: Blackwell Publishing, 2004), 380–412.
32 Clive Barnett, 'Geography and Ethics: Justice Unbound', *Progress in Human Geography* **35** (2), 246–55 (2011).
33 Steve Caney, 'Climate Change and the Duties of the Advantaged', *Critical Review of International Social and Political Philosophy* **13**, 203–28 (2010).
34 Julian Agyeman, 'Constructing Environmental (in)Justice: Transatlantic Tales', *Environmental Politics* **11** (3), 31–53 (2002).
35 Andre Szasz, *Ecopopulism: Toxic Waste and the Movement for Environmental Justice* (University of Minnesota Press, Minneapolis, 1994).
36 Chukwumerije Okereke, 'Global Environmental Sustainability: Intragenerational Equity and Conceptions of Justice in Multilateral Environmental Regimes', *Geoforum* **37** (5), 725–38 (2006).
37 Chukwumerije Okereke and Kate Dooley. 'Principles of Justice in Proposals and Policy Approaches to Avoided Deforestation: Towards a Post-Kyoto Climate Agreement', *Global Environmental Change* **20** (1), 82–95 (2010).
38 See Schlosberg, 'Theorising Environmental Justice', 37–55; see also David Schlosberg and David Carruthers, 'Indigenous Struggles, Environmental Justice, and Community Capabilities', *Global Environmental Politics* **10** (4), 12–35 (2010).
39 Adrian Martin, Nicole Gross-Camp, Bekeret Kebede, Shawn McGuire and Joseph Munyarukaza, 'Whose environmental justice? Exploring local and global perspectives in a payment for ecosystem services scheme in Rwanda', *Geoforum* **54**, 2 (2013).
40 Martin et al., 'Whose environmental justice?', 10.
41 Walker, 'Beyond Distribution and Proximity', 622.
42 Schlosberg, 'Theorising Environmental Justice', 45.
43 Schlosberg, 'Theorising Environmental Justice', 37.
44 Amartya Sen, *The Idea of Justice* (London: Allen Lane, 2009).
45 Barnett, 'Geography and Ethics', 252.
46 Fraser, *Justice Interruptus*.
47 Schlosberg, 'Theorising Environmental Justice', 37–55.
48 Bulkeley et al., 'Climate Justice and Global Cities', 914–25.
49 Nancy Fraser, *Scales of Justice* (Cambridge: Polity Press, 2008).
50 Gordon Walker and Rosie Day, 'Fuel poverty as injustice: Integrating distribution, recognition and procedure in the struggle for affordable warmth', *Energy Policy* **49**, 69–75 (2012).
51 See Robert D. Bullard, 'Dismantling Environmental Racism in the USA', *Local Environment* **4** (1), 5–19 (1999); see also Liere Urkidi and Mariana Walter, 'Dimensions of Environmental Justice in Anti-gold Mining Movements in Latin America', *Geoforum* **42**, 683–95 (2011).

52 Dorceta E. Taylor, 'The rise of the environmental justice paradigm: injustice framing and the social construction of environmental discourses', *American Behavioral Scientist* **43** (4), 508–80 (2000).
53 See Julian Agyeman, Robert D. Bullard and Bob Evans, *Just Sustainabilities: Development in an Unequal World* (Cambridge, MA: MIT Press, 2003); see also Julian Agyeman and Bob Evans, '"Just Sustainability": The Emerging Discourse of Environmental Justice in Britain?', *The Geographical Journal* **170** (2), 155–64 (2004); Julian Agyeman, *Sustainable Communities and the Challenge of Environmental Justice* (New York: New York University Press, 2005).
54 Agyeman, 'Constructing Environmental (in)Justice', 37.
55 Jane I. Dawson, 'The Two Faces of Environmental Justice: Lessons from the eco-nationalist Phenomenon', *Environmental Politics* **9** (2), 22–60 (2000).
56 Julia Miller Cantzler, 'Environmental Justice and Social Power Rhetoric in the Moral Battle over Whaling', *Sociological Inquiry* **77** (3), 483–512 (2007).
57 Chi, 'Capitalist Expansion and Indigenous Land Rights', 135–53.
58 See McDonald,. *Environmental Justice in South Africa*; see also Visser, 'Spatialities of social justice and local government transition', 99–111.
59 Dawson, 'The Two Faces of Environmental Justice', 36.
60 David Naguib Pellow and Robert J. Brulle. 'Power, Justice, and the Environment: Toward Critical Environmental Justice Studies', *Power, Justice, and the Environment: A Critical Appraisal of the Environmental Justice Movement* (Cambridge, MA: MIT Press, 2005), 1–19.
61 Kersty Hobson, 'Enacting Environmental Justice in Singapore: Performative Justice and the Green Volunteer Network', *Geoforum* **37** (5), 671–81 (2006).
62 Sam Barrett, 'The Necessity of a Multiscalar Analysis of Climate Justice', *Progress in Human Geography* **37** (2), 215–33 (2013).

Chapter 9

1 Jason Dittmer, Sami Moisio, Alan Ingram and Klaus Dodds, 'Have You Heard the One about the Disappearing Ice? Recasting Arctic Geopolitics', *Political Geography* **30** (4), 202–14 (2011); Elizabeth Nyman, 'Understanding the Arctic: Three Popular Media Views on the North', *Political Geography* **31** (6), 399–401 (2012).
2 Øistein Harsem, Arne Eide and Knut Heen, 'Factors Influencing Future Oil and Gas Prospects in the Arctic', *Energy Policy* **39** (12), 8037–45 (2011).
3 Nong Hong, 'The Energy Factor in the Arctic Dispute: A Pathway to Conflict or Cooperation?', *The Journal of World Energy Law & Business* **5** (1), 13–26 (2012).
4 Kathrin Keil, 'The Arctic: A New Region of Conflict? The Case of Oil and Gas', *Cooperation and Conflict* **49** (2), 162–90 (2014).
5 Graça Ermida, 'Energy Outlook for the Arctic: 2020 and beyond', *Polar Record* **52** (2), 170–5 (2016).
6 Kolson Schlosser, 'History, Scale and the Political Ecology of Ethical Diamonds in Kugluktuk, Nunavut', *Journal of Political Ecology* **20**, 53–69 (2013).
7 Jørgen Taagholt and Kent Brooks, 'Mineral Riches: A Route to Greenland's Independence?', *Polar Record* **52** (3), 360–71 (2016).
8 Keith Barney, 'Laos and the making of a "relational" resource frontier', *The Geographical Journal* **175** (2), 146–59 (2009).
9 Barney, 'Laos and the making of a "relational" resource frontier'.
10 James Van Alstine, Jacob Manyindo, Laura Smith, Jami Dixon and Ivan AmanigaRuhanga, 'Resource governance dynamics: The challenge of "new oil" in Uganda', *Resources Policy* **40**, 48–58 (2014).
11 Richard M. Auty, *Sustaining Development in Mineral Economies: The Resource Curse Thesis* (London: Routledge, 1993); Michael Ross, 'The Political Economy of the Resource Curse', *World Politics* **51**, 297–322 (1999); Jeffrey D. Sachs and Andrew D. Warner, 'Natural Resources and Economic Development: The Curse of Natural Resources', *European Economic Review* **45**, 827–38 (2001).
12 Van Alstine et al.,'Resource governance dynamics'.

13 Halvor Mehlum, Karl Moene, and Ragnar Torvik, 'Institutions and the Resource Curse', *The Economic Journal* **116**, 1–20 (2006); James A. Robinson, Ragnar Torvik and Thierry Verdier, 'Political foundations of the resource curse', *Journal of Development Economics* **79**, 447–68 (2006).
14 Timo Koivurova, 'Limits and Possibilities of the Arctic Council in a Rapidly Changing Scene of Arctic Governance', *Polar Record* **46** (2), 146–56 (2010); Siri Veland and Amanda H. Lynch, 'Arctic Ice Edge Narratives: Scale, Discourse and Ontological Security', *Area* **49**, 9–17 (2017).
15 Gavin Bridge and Philippe Le Billon, *Oil* (Cambridge: Polity, 2012).
16 Arctic Monitoring and Assessment (AMAP), *Arctic Oil and Gas 2007*. Available at www.amap.no.
17 Maria Ackrén and Bjarne Lindström, 'Autonomy development, irredentism and secessionism in a Nordic context', *Commonwealth & Comparative Politics*, **50** (4), 494–511 (2012).
18 Tasha R. Wyatt, 'Atuarfitsialak: Greenland's cultural compatible reform', *International Journal of Qualitative Studies in Education* **25** (6), 819–36 (2012).
19 Mark Nuttall, 'Self-Rule in Greenland: Towards the World's First Independent Inuit State?', *Indigenous Affairs* 3–4(08) (2008).
20 Harsem et al., 'Factors Influencing Future Oil and Gas Prospects in the Arctic'
21 Nuttall, 'Self-Rule in Greenland: Towards the World's First Independent Inuit State?'
22 Nuttall, 'Self-Rule in Greenland: Towards the World's First Independent Inuit State?'
23 Nuttall, Self-Rule in Greenland: Towards the World's First Independent Inuit State?'
24 L. Ford, 'Uganda vows to "defeat these thieves" in bid to reassure aid donors', *The Guardian* (2012).
25 World Bank, *The World Bank Indicators* (Washington, DC: The World Bank, 2016). Available at: http://data.worldbank.org/indicator/SI.POV.NAHC?locations=UG
26 CIA, *The World Factbook: Uganda* (Washington, DC: Central Intelligence Agency, 2016). Available at: https://www.cia.gov/library/publications/the-world-factbook/geos/ug.html
27 PWYP, Publish What You Pay Uganda Webpage (London: Publish What You Pay, 2016). Available at: http://www.publishwhatyoupay.org/members/uganda/
28 Paul Collier, *The Bottom Billion* (Oxford: Oxford University Press, 2007).
29 Graham McDowell and James D. Ford, 'The Socio-Ecological Dimensions of Hydrocarbon Development in the Disko Bay Region of Greenland: Opportunities, Risks, and Tradeoffs', *Applied Geography* **46** (January), 98–110 (2014).
30 Jon Rytter Hasle, Urban Kjellén and Ole Haugerud, 'Decision on Oil and Gas Exploration in an Arctic Area: Case Study from the Norwegian Barents Sea', *Safety Science* **47** (6), 832–42 (2009).
31 USGS, '90 Billion Barrels of Oil and 1,670 Trillion Cubic Feet of Natural Gas Assessed in the Arctic', US Department of the Interior, US Geological Survey (2008).
32 Lars Lindholt, and Solveig Glomsrød, 'The Arctic: No Big Bonanza for the Global Petroleum Industry', *Energy Economics* **34** (5), 1465–74 (2012).
33 Mars Nyvold, 'Low Price of Oil Puts Paid to Oil Adventure', *Greenland Oil & Minerals* (2016).
34 Bridge and Le Billon, *Oil*.
35 Oran R. Young, *Arctic Politics: Conflict and Cooperation in the Circumpolar North* (Lebanon, NH: Dartmouth College Press, 1992).
36 Van Alstine et al., 'Resource governance dynamics'.
37 Lassi Heininen, *Arctic Yearbook 2013* (Akureyri: Northern Research Forum, 2013).
38 Paul Arthur Berkman, Oran R. Young, et al., 'Governance and Environmental Change in the Arctic Ocean', *Science* **324** (5925), 339–40 (2009).
39 Timo Koivurova, 'Environmental Protection in the Arctic and Antarctic: Can the Polar Regimes Learn from Each Other?', *International Journal of Legal Information* **33** (2), 204–18 (2005).
40 Annika E. Nilsson, 'The Arctic Environment – From Low to High Politics', in Lassi Heininen (ed.), *Arctic Yearbook 2012* (Akureyri, Northern Research Forum, 2012), 180–94.
41 Hannes Gerhardt, Philip E. Steinberg, Jeremy Tasch, Sandra J. Fabiano and Rob Shields, 'Contested Sovereignty in a Changing Arctic', *Annals of the Association of American Geographers* **100** (4), 992–1002 (2010).
42 Nilsson, 'The Arctic Environment'.

43 Marieme Jamme, 'Negative perceptions slow Africa's development', *The Guardian*. Global Development, Poverty Matters Blog. 10 December 2010. Available at: https://www.theguardian.com/global-development/poverty-matters/2010/dec/10/africa-postcolonial-perceptions
44 Oleg A. Anisimov, David G. Vaughan, T. V. Callaghan, Christopher Furgal, Harvey Marchant, Terry D. Prowse, Hjalmar Vilhjálmsson and John E. Walsh, 'Polar regions (Arctic and Antarctic)', in M. L. Parry, O. F. Canziani, J. P. Palutikof, P. J. van der Linden and C. E. Hanson (eds), *Climate Change 2007: Impacts, Adaptation and Vulnerability. Contribution of Working Group II to the Fourth Assessment Report of the Intergovernmental Panel on Climate Change* (Cambridge: Cambridge University Press, 2007), 653–85.
45 Terry V. Callaghan, Lars Olof Björn, F. S. Chapin III, Y. Chernov, Torben R. Christensen, Brian Huntley, Rolf Ims, et al., 'Arctic Tundra and Polar Desert Ecosystems', in Carolyn Symon, Lelani Arris and Bill Heal (eds), *Arctic Climate Impact Assessment* (New York: Cambridge University Press, 2005), 243–352.
46 Anisimov et al., 'Polar regions (Arctic and Antarctic)'.
47 Jason P. Briner, 'Climate Science: Ice Streams Waned as Ice Sheets Shrank', *Nature* **530** (7590), 287–8 (2016).
48 William Davies, James Van Alstine and Jon C. Lovett, 'Frame Conflicts in Natural Resource Use: Exploring Framings around Arctic Offshore Petroleum Using Q-Methodology', *Environmental Policy and Governance* (2016). Accessed 15 August 2016, doi:10.1002/eet.1729
49 IPCC, *Climate Change 2014: Synthesis Report. Contribution of Working Groups I, II and III to the Fifth Assessment Report of the Intergovernmental Panel on Climate Change* [Core Writing Team, R. K. Pachauri and L. A. Meyer (eds)]. (Geneva: IPCC, 2014).
50 EAC, *EAC Climate Change Policy* (Arusha, Tanzania: East African Community Secretariat, 2011).
51 EAC, *EAC Climate Change Policy.*
52 Shelagh Whitley and Godber Tumushabe, *Mapping Current Incentives and Investment in Uganda's Energy Sector* (London: ODI and Kampala: ACODE, 2014).
53 Whitley and Tumushabe, *Mapping Current Incentives and Investment in Uganda's Energy Sector.*
54 James Van Alstine, 'Transparency in Resource Governance: The Pitfalls and Potential of "New Oil" in Sub-Saharan Africa', *Global Environmental Politics* **14** (1), 20–39 (2014).
55 Coco C. A. Smits, Jan P. M. van Tatenhove and Judith van Leeuwen, 'Authority in Arctic Governance: Changing Spheres of Authority in Greenlandic Offshore Oil and Gas Developments', *International Environmental Agreements: Politics, Law and Economics* **14** (4), 329–48 (2014).
56 Smits et al., 'Authority in Arctic Governance'.
57 Cécile Pelaudeix, 'Governance of Arctic Offshore Oil & Gas Activities: Multilevel Governance & Legal Pluralism at Stake', in Lassi Heininen, Heather Exner-Pirot and Joël Plouffe (eds), *Arctic Yearbook 2015* (Akureyri: Northern Research Forum, 2015), 214–33.
58 Juha Käpylä and Harri Mikkola, 'On Arctic Exceptionalism: Critical Reflections in the Light of the Arctic Sunrise Case and the Crisis in Ukraine', The Finnish Institute of International Affairs. Working Paper 85 (2015).
59 Susanah Stoessel, Elizabeth Tedsen, Sandra Cavalieri and Arne Riedel, 'Environmental Governance in the Marine Arctic', in Elizabeth Tedsen, Sandra Cavalieri and R. Andreas Kraemer (eds), *Arctic Marine Governance: Opportunities for Transatlantic Cooperation* (New York: Springer, 2013), 45–69.
60 Smits et al., 'Authority in Arctic Governance'.
61 Pelaudeix, 'Governance of Arctic Offshore Oil & Gas Activities'.
62 African Union, 'AU in a nutshell', African Union Website (2016). Accessed 25 November 2016, http://au.int/en/about/nutshell
63 Republic of Uganda. *Uganda Vision 2040* (Kampala: Government of Uganda, 2013), p. xiii. Available at: http://npa.ug/uganda-vision-2040
64 KPMG, *East Africa Regional Cooperation in Oil and Gas: Possible Reality?* (Nairobi: KPMG East Africa, 2016).
65 KPMG, *East Africa Regional Cooperation in Oil and Gas.*
66 Peter G. Veit, Carole Excell and Alisa Zomer, 'Avoiding the Resource Curse: Spotlight on oil in Uganda', WRI Working Paper, World Resources Institute, Washington, DC, 2011.

67 Emma Wilson and James Van Alstine, *Localising Transparency: Exploring EITI's Contribution to Sustainable Development* (London: International Institute for Environment and Development, 2014).
68 The Monitor. 'Oil – A Dream Come True for Museveni', *The Monitor.* Africa News, Uganda (2006), no page.
69 Young, *Arctic Politics.*
70 Mark Nuttall, 'Imagining and Governing the Greenlandic Resource Frontier', *The Polar Journal* **2** (1), 113–24 (2012).
71 McDowell and Ford, 'The Socio-Ecological Dimensions of Hydrocarbon Development in the Disko Bay Region of Greenland'; William Davies, Samuel Wright and James Van Alstine, 'Framing a "Climate Change Frontier": International News Media Coverage Surrounding Natural Resource Development in Greenland', *Environmental Values*, in press; Emma Wilson, *Energy and Minerals in Greenland: Governance, Corporate Responsibility and Social Resilience* (London: International Institute for Environment and Development, 2015).
72 McDowell and Ford, 'The Socio-Ecological Dimensions of Hydrocarbon Development in the Disko Bay Region of Greenland'.
73 Wilson, *Energy and Minerals in Greenland.*
74 Davies et al., 'Framing a "Climate Change Frontier"'.
75 Jacob Manyindo, James Van Alstine, Ivan Amaniga Ruhanga, Emanuel Mukuru, Laura Smith, Christine Nantongo and Jen Dyer, *The Governance of Hydrocarbons in Uganda: Creating Opportunities for Multi-Stakeholder Engagement* (Kampala: Maendeleo ya Jamii, 2014).
76 James Van Alstine and Stavros Afionis, 'Community and company capacity: the challenge of resource-led development in Zambia's "New Copperbelt"', *Community Development Journal* **48** (3), 360–76 (2013).
77 Lisa Gregoire, 'Greenland pushing ahead with oil and gas development', Nunatsiaq Online, 15 May 2014, accessed 20 July 2016, http://www.nunatsiaqonline.ca/stories/article/65674greenland_pushing_ahead_with_oil_and_gas_development
78 Kevin McGwin, 'Not necessarily opposed, just sceptical', *Arctic Journal* 14 May 2014, accessed 20 July 2016, http://arcticjournal.com/climate/611/not-necessarily-opposed-just-sceptical
79 Gregoire, 'Greenland pushing ahead with oil and gas development'.

Chapter 10

1 See Monica Tennberg, 'National and Regional Approaches to Arctic Adaptation Governance', in Timo Koivurova, E. Karina, H. Keskitalo and Nigel Bankes (eds), *Climate Governance in the Arctic* (Heidelberg: Springer, 2009), 289–301, at 299.
2 For instance, the Chapeau of the International Convention for the Regulation of Whaling (ICRW) reads: 'The Governments … Having decided to conclude a convention to provide for the proper conservation of whale stocks and thus make possible the orderly development of the whaling industry' (the International Convention for the Regulation of Whaling, 2 December 1946 (161 UNTS 72)).
3 The full text of the Arbitration Treaty is reproduced in Cairo A. R. Robb (ed.), *International Environmental Law Reports, Vol. 1. Early Decisions* (Cambridge: Cambridge University Press, 1999).
4 Nikolas Sellheim, 'Early Arctic sealing regimes: the Bering Sea fur seal regime *vis-à-vis* Finnish–Soviet fishing and sealing agreements', *Polar Record* **52** (1), 109–14 (2016), at 109.
5 Oran R. Young, *Institutional Dynamics: Emergent Patterns in International Environmental Governance* (Cambridge, MA: MIT Press, 2010).
6 See Robert L. Friedheim (ed.), *Toward a Sustainable Whaling Regime* (Seattle: University of Washington Press, 2001).
7 Since 1982 a moratorium on commercial whaling has been in place.
8 Malgosia Fitzmaurice, *Whaling and International Law* (Cambridge: Cambridge University Press, 2016), 248.
9 Regulation (EC) No 1007/2009 on Trade in Seal products ('Basic Regulation') and Commission Regulation (EU) laying down Detailed Rules for the Implementation of Regulation (EC) No 1007/2009 ('Implementing Regulation'). On the drafting process, see

Ferdi de Ville, 'Explaining the genesis of a trade dispute: The European Union's seal trade ban', *Journal of European Integration* **34** (1), 37–53 (2012).

10 Katie Sykes, 'Sealing animal welfare into the GATT exceptions: the international dimension of animal welfare in WTO disputes', *World Trade Review* **13** (3), 471–98 (2014).

11 Although these requirements are to be monitored by a 'Recognised body', it is unclear how the animal welfare requirements in particular will be implemented. After all, hunters in the Canadian and Greenlandic Arctic conduct seal hunting in a very remote area and little monitoring – if any – is possible.

12 Basic Regulation, Article 3(b), as amended.

13 Nikolas Sellheim, 'The Narrated "Other" – Challenging Inuit Sustainability through the European Discourse on the Seal Hunt', in Kamrul Hossain and Anna Petrétei (eds), *Understanding the Many Faces of Human Security. Perspectives of Northern Indigenous Peoples* (Leiden: Brill, 2016), 56–73, at 65, 66.

14 See also Dorothée Cambou, 'The Impact of the Ban on Seal Products on the Rights of Indigenous Peoples: A European Issue', *The Yearbook of Polar Law* **V**, 389–415, (2013); Nikolas Sellheim, '"Direct and Individual Concern" for Newfoundland's Sealing Industry? When a Legal Concept and Empirical Data Collide', *The Yearbook of Polar Law* **VI**, 466–96 (2015).

15 See also Mark Nuttall and Terry V. Callaghan, 'Introduction', in Mark Nuttall and Terry V. Callaghan (eds), *The Arctic: Environment, People, Policy* (Boca Raton: CRC Press, 2000), xxv–xxxviii, at xxix–xxxii.

16 The question of a definition of culture is subject to ongoing scrutiny and an extremely rich body of literature exists on the issue. Suffice it to say that the question 'What is culture?' is heatedly debated.

17 Solveig Glømsrod and Iulie Aslaksen (eds), *The Economy of the North 2008* (Oslo: Statistics Norway, 2009).

18 Convention on Biological Diversity, 5 June 1992 (1760 UNTS 79).

19 Oceans Act, S. C. 1996, c. 31, 18 December 1996.

20 Andrea H. Procter, 'Konstruktion von Machstrukturen und traditionellem Wissen im nördlichen Kanada' ['Construction of power structures and traditional knowledge in northern Canada'] in Stefan Bauer, Stefan Donecker, Aline Ehrenfried and Markus Hirnsperger (eds), *Bruchlinien im Eis. Ethnologie des zirkumpolaren Nordens* [*Break lines in the ice. Ethnology of the Circumpolar North*] (Münster: LIT Verlag, 2005), 208–19.

21 The body of knowledge, technology and institutions held by the Inuit is reflected in the Inuktitut phrase *Inuit Qaujimajatuqangit*.

22 Own emphasis; Fikret Berkes, *Sacred Ecology. Traditional Ecological Knowledge and Resource Management* (Philadelphia: Taylor & Francis, 1999), 130.

23 This is best exemplified in Clifford Geertz's seminal work *Local Knowledge* (London: Fontana Press, 1993). On the emergence of 'new' traditional systems, see for example Shaun Larcom, *Legal Dissonance. The Interaction of Criminal Law and Customary Law in Papua New Guinea* (New York & Oxford: Berghahn Books, 2015).

24 'Museification' describes the perception of a culture (typically indigenous) by outsiders putting it in a box or a museum, making it unchanging, less developed than one's own, simplified and therefore without connection to the real world; in effect seeing a culture how one would like it to be seen.

25 The history of colonisation of Lapland gives profound insight into the way cultures have, or were, merged up to this day. While primarily the settlers imposed their culture on the native Saami (see for example Marja Tuominen, 'History as a Project of Progress? The North as a Focus of Attention', in Aini Linjakumpu and Sandra Wallenius-Korkalo (eds), *Progress or Perish. Northern Perspectives on Social Change* (London & New York: Routledge, 2016), 11–32); also Saami culture, particularly language, is now reflected in Scandinavian society (Jurij K. Kusmenko, *Der samische Einfluss auf die skandinavischen Sprachen* [*The Saami Influence on the Scandinavian Languages*] (Berlin: Nordeuropa Institut, 2011).

26 International Covenant for Civil and Political Rights (ICCPR) and International Covenant for Economic, Social and Cultural Rights (ICESCR), 1966. The case was launched by Saami reindeer herders in response to Finland's plans to license stone quarrying in Saami reindeer herding areas, who saw their rights violated under ICCPR article 27, which protects the cultures of minorities.

27 Original emphasis; *Länsman et al. v Finland*, Communication No. 511/1992, U.N. Doc. CCPR/C/52/D/511/1992 (1994), para. 9.3.

28 See Stephen Gudeman, *The Anthropology of Economy: Community, Market and Culture* (Hoboken: Wiley-Blackwell, 2001). The perception of Arctic cultures and economies in a somewhat outdated fashion is not the only component of the Arctic which is approached in this manner. Craciun shows how exploration and geography are still looked at through a nineteenth-century lens (Adriana Craciun, *Writing Arctic Disaster. Authorship and Exploration* (Cambridge: Cambridge University Press, 2016)).
29 Nikolas Sellheim, 'The Neglected Tradition? The Genesis of the EU Seal Product Trade Ban and Commercial Sealing', *The Yearbook of Polar Law* V, 417–50 (2013).
30 Natalia Loukacheva, *Arctic Justice. Legal and Political Autonomy of Greenland and Nunavut* (Toronto: University of Toronto Press, 2007), 87–91.
31 Nikolas Sellheim, 'Seal Hunting in the Arctic States. An Analysis of Legislative Frameworks, Incentives and Histories', *The Yearbook of Polar Law* VII, 188–224 (2015).
32 Greenland's external relations are represented by Denmark.
33 Declarations in EU law do not have legally binding effects, but carry political weight.
34 The Treaty of Lisbon does not have Declaration 25 annexed to it. Whether or not it is still valid is interpreted differently (see Fitzmaurice, *Whaling*, 213, 214).
35 See also International Whaling Commission. *Infractions Sub-Committee. Summary of the Main Outcomes*. IWC/66/Rep04. Accessed 3 November 2016. file:///C:/Users/The%20 Awesomes/Downloads/RS6343_66-Rep04.pdf
36 ACIA, *Arctic Climate Impact Assessment. Impacts of a Warming Arctic* (Cambridge: Cambridge University Press, 2005).
37 Meinhard Doelle, 'The Climate Change Regime and the Arctic Region', in Koivurova et al. (eds), *Climate Governance in the Arctic*, 27–50.
38 See for instance Kirsten Hastrup and Maria Louise B. Robertson, 'Fixed and fluid waters – mirroring the Arctic and the Pacific', in Kirsten Hastrup and Cecilie Rubow (eds), *Living with Environmental Change – Waterworlds* (New York & London: Routledge, 2014), 80–7.
39 Frank Sejersen, *Rethinking Greenland and the Arctic in the Era of Climate Change* (London & New York: Routledge, 2015).
40 For a comprehensive analysis, see Md Waliul Hasanat, *Soft-law Cooperation in International Law. The Arctic Council's Efforts to Address Climate Change*. Doctoral dissertation. Acta Universitatis Lapponiensis 234 (Rovaniemi: Rovaniemi University Press, 2012).
41 Rune Rafaelsen, 'The Barents Cooperation – New Regional Approach for Foreign Policy in the High North', in Atle Staalesen (ed.), *Talking Barents – People, Borders and Regional Cooperation* (Kirkenes: Norwegian Barents Secretariat, 2010), 25–31.
42 Wilfried Greaves, 'Environment, Identity, Autonomy: Inuit Perspectives on Arctic Security', in Hossain and Petrétei (eds), *Understanding the Many Faces of Human Security*, 35–55.

Chapter 11

1 Robert W. Murray and Anita Dey Nuttal, 'Understanding Policy and Governance in the Arctic', in Robert W. Murray and Anita Dey Nuttal (eds), *International Relations and the Arctic: Understanding Policy and Governance* (New York: Cambria Press, 2014), 1.
2 'Anxiously Watching a Different World', *The Economist*, May 2007, accessed November 2016 http://www.economist.com/node/9225715
3 Mieke Coppes and Victoria Hermann, 'First Past the Post: Harper, Trudeau, and Canada's Arctic Values', *The Arctic Institute*, October 2015, accessed November 2016 http://www.thearcticinstitute.org/first-past-the-pole-harper-trudeau/
4 Greg Sharp and Andreas Østhagan, 'A New Canadian Government ... So What?', *The Arctic Institute*, June 2016, accessed November 2016 http://www.thearcticinstitute.org/a-new-canadian-government-so-what/
5 'Anxiously Watching a Different World'.
6 United Nations, 'United Nations Declaration on the Rights of Indigenous Peoples', March 2008, accessed November 2016, http://www.un.org/esa/socdev/unpfii/documents/DRIPS_en.pdf
7 Government of Yukon, 'Population Report: Second Quarter 2016', accessed November 2016, http://www.eco.gov.yk.ca/pdf/populationJun_2016.pdf
8 AKCanada, 'Living in Canada: The Northwest Territories', accessed November 2016, https://www.akcanada.com/lic_northwestterritories.php

9 Yukon Bureau of Statistics, 'Immigration and Ethnocultural Diverstiy: 2011 National Household Survey', January 2014, accessed November 2016, http://www.eco.gov.yk.ca/pdf/Immigration_and_Ethnocultural_Diversity_2011.pdf
10 Yukon Bureau of Statistics, 'Immigration and Ethnocultural Diverstiy: 2011 National Household Survey', January 2014, accessed November 2016, http://www.eco.gov.yk.ca/pdf/Immigration_and_Ethnocultural_Diversity_2011.pdf
11 Government of the Northwest Territories, 'Employers in the Northwest Territories', accessed November 2016, http://www.immigratenwt.ca/employers-northwest-territories
12 Jackson Lafferty, 'Enhancements to the NWT Nominee Programme', accessed November 2016, http://www.gov.nt.ca/newsroom/jackson-lafferty-enhancements-nwt-nominee-program.
13 Northwest Territories Language Commissioner, 'NWT Official Languages', accessed November 2016, http://www.nwtlanguagescommissioner.ca/nwt-official-languages/
14 Truth and Reconciliation Commission, 'Canada's Residential Schools: The Inuit and Northern Experience: The Final Report of the Truth and Reconciliation Commission of Canada Volume 2' (Montreal: McGill and Queen's Press, 2015), accessed November 2016, http://www.myrobust.com/websites/trcinstitution/File/Reports/Volume_2_Inuit_and_Northern_English_Web.pdf
15 Truth and Reconciliation Commission, 'Canada's Residential Schools: Volume 2', 73–4.
16 Truth and Reconciliation Commission, 'Canada's Residential Schools: Volume 2', 187.
17 Truth and Reconciliation Commission, 'Canada's Residential Schools: Volume 2', 187
18 Ann Vick-Westgate, *Nunavik: Inuit-controlled Education in Arctic Quebec* (Calgary: University of Calgary Press, 2002), xviii.
19 Westgate, *Nunavik*, ix.
20 Heather E. McGregor, *Inuit Education and Schools in the Eastern Arctic* (Vancouver: UBC Press, 2010), xi.
21 Franklyn Griffiths, 'The Arctic in the Russian identity', *The Soviet Maritime Arctic* 83–107, 86 (1991).
22 Y. P. Shabayev, Zherebtsov and P. S. Zhuravlyov, 'Russkiy Sever': kulturnye i kulturnye smysly. *Mir Rossii* No. 4 (2012) (in Russian).
23 V. N. Kalutskov, 'Kulturno-geographicheskoye rayonirovaniye Rossii: geokontseptualny podhod', *Pskovskiy Regionalniy Journal* No 22 (2015) (in Russian).
24 See http://fom.ru/Mir/12216
25 Federalny Zakon Rossiyskoy Federatsii, 'O strategicheskom planirovanii v Rossiyskoy Federatsii', 28 iunya 2014, No. 172-ФЗ. Available at: http://www.rg.ru/2014/07/03/strategia-dok.html
26 V. P. Emelyantsev, 'Upravleniye razvitiem makroregionov (na primere Arkticheskoy territorii Dalnego Vostoka)', *Rossiyskaya Arktika – Territoriya prava* (Moscow, 2014), 19–47 (in Russian).
27 P. A. Minakir and A. P. Gorynov, 'Prostranstvenno-ekonomicheskie aspekty osvoyeniya Arktiki', *Vestnik MGU* No. 3 (2015) (in Russian).
28 See e.g. H. A. Conley and C. Rohloff, *The New Ice Curtain: Russia's Strategic Reach to the Arctic* (Lanham, MD: Rowman & Littlefield, 2015).
29 D. Carr, *Time, Narrative, and History* (Bloomington: Indiana University Press; 1986), 61.
30 Prezident Rossiyskoy Federatsii, 'Osnovy gosudarstvennoy politiki Rossiyskoy Federatsii v Arktike na period do 2020 goda i na dalneyshuyu perspektivu'. 18 sent. 2008, Пр – 1969; Prezident Rossiyskoy Federatsii, 'Strategiya razvitiya Arkticheskoy zony Rossiyskoy Federatsii i obespecheniya natsionalnoy bezopasnosti na period do 2020 goda'. 8 fevr. 2013, Пр-232. The English translation of the Foundations of the state policy of the Russian Federation in the Arctic for the period until 2020 and beyond can be found at http://icr.arcticportal.org/index.php?option=com_content&view=article&id=1791%3
31 Marlene Laruelle, *Russia's Arctic Strategies and the Future of the Far North* (Armonk, NY: ME Sharpe, 2013).
32 V. E. Seliverstov, 'Federalism, regional growth, and regional science in post-Soviet Russia: Modernization or degradation?', *Regional Research of Russia* **4** (4), 240–52.
33 John McCannon, *Red Arctic: Polar Exploration and the Myth of the North in the Soviet Union, 1932–1939* (New York: Oxford University Press, 1998).
34 E.g. John Brian Harley and Paul Laxton, *The New Nature of Maps: Essays in the History of Cartography*. No. 2002 (Baltimore, MD: Johns Hopkins University Press, 2002).
35 Ivan Ivanovich Semyonov, head of Anabarskiy national ulus (district), Sakha Republic (Yakutia), since 2013.

36 N. Thompson, *Settlers on the Edge: Identity and Modernisation on Russia's Farthest Arctic Frontier* (Vancouver and Seattle: University of British Columbia Press and University of Washington Press, 2008), 9.
37 Quoted from http://www.tv21.ru/news/2016/10/31/stolice-severa--s-lyubovyu-v-murmanske-vystupila-nadezhda-babkina)
38 E.g. A. V. Kozlov, S. V. Fedoseev and A. E. Cherepovitsyn, *Kompleksnoye razvitiye ekonomicheskogo prostranstva Arkticheskoy zony Rossiysky Federatsii* (SPb., Izd-vo Politechn. Un-ta., 2016).
39 Quoted from http://www.arctic-info.ru/news/15-12-2015/eksperti--arktike-nyjni-desatki-tisac-specialistov/
40 The list of Small-numbered Indigenous Peoples of the North, Siberia and the Far East was updated in 2015 and published in the Russian newspaper (Pravitelstvo Rossiyskoy Federatsii, 'Postanovleniye O yedinom perechne korennyh malochislennyh narodov Severa, Sibiri i Dalnego Vostoka', 24.03.2000 No.255 (ed. 25.08.2015). Available at: https://rg.ru/2015/10/12/sever-dok.html)
41 Sovyet Ministrov RSFSR, 'Postanovleniye 'Ob okazanii dopolnitelnoy pomoschi v razvitii hozyaystva i kultury narodov Severa', in Systemnoye sobranie zakonov RSFR, ukazov prezidiuma Verhovnogo Sovieta RSFSR i razresheniy pravitelstva RSFSR. T. 8 (Moscow, 1960).
42 F. H. Sokolova, 'Korennye malochislennye narody Arktiki'. AiS 2013, No 12. Available at: http://cyberleninka.ru/article/n/korennye-malochislennye-narody-arktiki-kontsept-sovremennoe-sostoyanie-kultury
43 Pravitelstvo Rossiyskoy Federatsii, 'Rasporyazheniye dated 25 of August 2016'. Available at: http://government.ru/media/files/680AUxeuDCVBP2RSPmo1VoGoWKTDUqf5.pdf
44 Pravitelstvo Rossiyskoy Federatsii, 'Kontseptsia ustoychivogo razvitiya korennyh malochislennyh narodov Seevera, Sibiri i Dalnego Vostoka'. Utv. Rasporyazheniem Pravitelstva RF dated 4 February 2009, No. 132-p. Available at: http://gov.garant.ru/document?id=94908&byPara=1&sub=1
45 Quoted from Departament po delam korennyh molochislennyh narodov Severa. 'Informatsiya or razvitii kochevogo obrazovaniya v Yamalo-Nenetskom Avtonomnom Okruge'. [online] (2013). Available at: http://dkmns.ru/kochevye-shkoly (in Russian).
46 A federal subject and territorial unit in Russia with a certain level of autonomy. Out of four AO, three are located within the Russian Arctic Zone.
47 See e.g. Ewa Thompson, 'It is Colonialism After All: Some Epistemological Remarks', *Teksty Drugie* **1**, 67–81 (2014). And the work by M. Khodarkovskiy, *Russia's Steppe Frontier: The Making of a Colonial Empire, 1500–1800* (Bloomington: Indiana University Press, 2002). And his interview in M. Sinness, 'Empire of the Steppe: Russia's Colonial Experience on the Eurasian Frontier'. [online] 2014. Available at: http://www.international.ucla.edu/euro/article/139315
48 See, for example, an interview with a former First Vice-President of the Russian Association of Indigenous Peoples of the North, Siberia and the Far East (RAIPON) and a Russian indigenous rights activist Pavel Sulyandziga who was removed from power in 2015 as a result of an internal feud (Y. Frolova, 'Pavel Sulyandziga: Vsyo, chto bylo v Salekharde, – eto istoriya'. Sluzhba Novostey URA-ru, 9 June 2016. Available at: http://ura.ru/articles/1036268108 (in Russian).
49 Deline Got'ine Government, 'About Us', accessed November 2016, http://deline.ca/en/about-us/
50 E.g. Alastair Pennycook, *English and the Discourses of Colonialism* (Abingdon: Routledge, 2002); Patrick Wolfe, *Settler Colonialism* (London: A&C Black, 1999).

Chapter 12

1 Kenneth Bird et al., '90 Billion barrels of oil and 1,670 Trillion Cubic feet of natural gas assessed in the Arctic', US Department of the Interior: US Geological Survey (USGS). 23 July 2008. Available online https://pubs.usgs.gov/fs/2008/3049/fs2008-3049.pdf
2 Lars Lindholdt, 'Arctic natural resources in a global perspective', in Solveig Glomsrød and Iulie Aslaksen (eds), *The Economy of the North* (Oslo, Norway: Statistics Norway, 2006), 27–37. http://www.ssb.no/a/english/publikasjoner/pdf/sa84_en/kap3.pdf

3 Greenpeace, *Russian Arctic: Offshore Hydrocarbon Exploration: Investment Risks* (Moscow: Greenpeace Russia, 2012). Accessed 6 November 2016. http://www.greenpeace.org/russia/Global/russia/report/Arctic-oil/ArcticSave_English_26_apr.pdf; Rune S. Fjellheim and John B. Henriksen, 'Oil and Gas Exploitation on Arctic Indigenous Peoples' Territories', *Gáldu Cála – Journal of Indigenous Peoples Rights* issue 4, 1–48 (2006).
4 Van Alstine and Davies, this volume.
5 Stephen Crowley, 'Monotowns and the political economy of industrial restructuring in Russia', *Post-Soviet Affairs* **32** (5), 397–422 (2016).
6 Trevor J. Barnes, 'Borderline communities: Canadian single industry towns, staples and Harold Innis', in Henk Van Houtum, Olivier Kramsch and Wolfgang Zierhofer (eds), *B/ordering Space*. (Farnham: Ashgate, 2005), 109–22; Michael Young, 'Help wanted: A call for the non-profit sector to increase services for hard-to-house persons with concurrent disorders in the Western Canadian Arctic', *The Extractive Industries and Society* **3**, 41–9 (2016).
7 Daniel Kempton and Terry Clark (eds), *Unity or Separation: Centre–periphery Relations in the Former Soviet Union* (London: Praeger, 2002).
8 Mattias Åhrén, *Indigenous Peoples' Status in the International Legal System* (Oxford: Oxford University Press, 2016).
9 Alberto Acosta, 'Extractivism and neoextractism: two sides of the same curse', in Miriam Lang and Dunia Mokrani (eds), *Beyond Development: Alternative Visions from Latin America* (Amsterdam: Transnational Institute, 2013), 61–87. Accessed 6 November 2016. https://www.tni.org/files/download/beyonddevelopment_complete.pdf
10 Florian Stammler and Aitalina Ivanova, 'Confrontation, coexistence or co-ignorance? negotiating resource rights in two Russian regions', *The Extractive Industries and Society* **3**, 60–71 (2016).
11 Lill Rastad Bjørst, 'Saving or destroying the local community? Conflicting spatial storylines in the Greenlandic debate on uranium', *The Extractive Industries and Society* **3**, 34–40 (2016).
12 Kathryn Oberdeck, 'Archives of the unbuilt environment: documents and discourses of imagined space in twentieth-century Kohler, Wisconsin', in A. Burton (ed.), *Archive Stories: Facts, fictions and the writing of history* (Chapel Hill, NC: University of North Carolina Press, 2006), 251–74; Jonathan Peyton, 'Corporate ecology: BC Hydro's Stikine-Iskut project and the unbuilt environment', *Journal of Historical Geography* **37**, 358–69 (2011).
13 Anne Merrild Hansen and Pelle Tejsner, 'Challenges and opportunities for residents in the Upernavik District while oil companies are making a first entrance in Baffin Bay', *Arctic Anthropology* **53** (1), 84–94 (2016); Anne Merrild Hansen, Frank Vanclay, Peter Croal and Anna-Sofie Hurup Skjervedal, 'Managing the social impacts of the rapidly-expanding extractive industries in Greenland', *The Extractive Industries and Society* **3** (1), 25–33 (2016).
14 Elana Wilson Rowe, 'Promises, Promises: Murmansk and the Unbuilt Petroleum Environment' (2017) Arctic Review on Law and Politics **8**, 3–16. http://dx.doi.org/10.23865/arctic.v8.504.
15 Emma Wilson, 'Landmark decision by Sami people to say no to gold mine in Norway', 11 November 2015. Accessed 6 November 2016. https://www.linkedin.com/pulse/landmark-decision-sami-people-say-gold-mine-norway-emma-wilson
16 Greenland Oil Industry Association, 'A summary of the oil & gas exploration history of Greenland' (2016). Accessed 9 November 2016. http://goia.gl/en-us/oilgasingreenland/history.aspx
17 Greenland Oil Industry Association, 'A summary of the oil & gas exploration history of Greenland'.
18 Greenland Oil Industry Association, 'A summary of the oil & gas exploration history of Greenland'.
19 Frank Sejersen, 'Resilience, human agency and climate change adaptation strategies in the Arctic', in T.R.D.A. o. S. a. Letters (ed.), *The Question of Resilience: Social responses to climate change* (Copenhagen: Det Kongelige Danske Videnskabernes Selskab, 2009), 218–43.
20 Pelle Tejsner, 'Quota disputes and subsistence whaling in Qeqertarsuaq, Greenland', *Polar Record* **50** (Special Issue 4), 430–9 (2014).
21 Government of Greenland, Selvstyrets bekendtgørelse nr. 7 af 29. marts 2011 om beskyttelse og fangst af hvid- og narhvaler (The Self Rule Government Law no. 7 of 29 March 2011

on protection and hunt of belugas and narwhales) (2011). Accessed 14 November 2016. http://arkiv.lovgivning.gl/gh.gl-love/dk/2011/context_2011.htm#Bekendtgørelser

22 Alyne Delaney, Rikke Becker Jacobsen and Kåre Hendriksen, 'Greenland Halibut in Upernavik: A preliminary study of the importance of the stock for the fishing populace'. Technical University of Denmark (2012). Accessed 14 November 2016. http://orbit.dtu.dk/files/10235133/Greenland_Halibut.pdf

23 National Environmental Research Institute, 'The eastern Baffin Bay. A preliminary strategic environmental impact assessment of hydrocarbon activities in the KANUMAS West area', edited by David Boertmann, Anders Mosbech, Doris Schiedek and Kasper Johansen (National Environmental Research Institute, Aarhus University, Denmark, 2009). National Environmental Research Institute Technical report no. 720.

24 Anna-Sofie H. Olsen and Anne M. Hansen, 'Stakeholder perceptions of public participation in Environmental Impact Assessment: A case study of offshore oil exploration industry in Northwest Greenland', *Impact Assessment and Project Appraisal* **32** (1), 72–80 (2014).

25 Personal communication, May 2014; cited in Hansen and Tejsner, 'Challenges and opportunities for residents in the Upernavik District'.

26 Lill Rastad Bjorst, 'Saving or destroying the local community? Conflicting spatial storylines in the Greenlandic debate on uranium', *The Extractive Industries and Society* **3**, 34–40 (2016); Mark Nuttall, 'Imagining and governing the Greenlandic resource frontier', *The Polar Journal* **2** (1), 113–24 (2012); M. Nuttall, 'Zero-tolerance, uranium and Greenland's mining future'. *The Polar Journal* **3** (2), 368–83 (2012); Emma Wilson, *Energy and Minerals in Greenland: Governance, Corporate Responsibility and Social Resilience* (London: International Institute for Environment and Development, 2015). Accessed 6 November 2016. http://pubs.iied.org/16561IIED/

27 Dag Harald Claes and Arild Moe, 'Arctic Petroleum Resources in a Regional and Global Perspective', in R. Tamnes and K. Offerdal (eds), *Geopolitics and Security in the Arctic: Regional Dynamics in a Global World* (Oxford: Routledge, 2014), 97–120.

28 Claes and Moe, 'Arctic Petroleum Resources in a Regional and Global Perspective'.

29 Wilson Rowe, 'Promises, Promises: Murmansk and the Unbuilt Petroleum Environment'.

30 Aileen Espiritu, 'Kautokeino and Kvalsund compared: rejection and acceptance of mining in communities in Northern Norway', *The Northern Review* **39**, 53–63 (2016); Vigdis Nygaard, 'Do indigenous interests have a say in planning of new mining projects? Experiences from Finnmark, Norway', *The Extractive Industries and Society* **3**, 17–24 (2016).

31 Statistics Norway, accessed 6 November 2016 https://www.ssb.no/186162/urban-settlements.population-and-area-by-municipality.1-january-2013

32 Nygaard, 'Do indigenous interests have a say in planning of new mining projects?'

33 Espiritu, 'Kautokeino and Kvalsund compared'.

34 Ellen Inga Turi and E. Carina H. Keskitalo, 'Governing reindeer husbandry in Western Finnmark: Barriers for incorporating traditional knowledge in local-level policy implementation', *Polar Geography*, **37** (3), 234–251 (2014).

35 Nygaard, 'Do indigenous interests have a say in planning of new mining projects?'

36 Kathrin Ivsett Johnsen, Tor A. Benjaminsen and Inger Marie Gaup Eira, 'Seeing like the state or like pastoralists? Conflicting narratives on the governance of Sami reindeer husbandry in Finnmark, Norway', *Norwegian Journal of Geography* **69** (4), 230–41 (2015).

37 Personal communication, business leader, Kautokeino, 23 September 2015.

38 Personal communication, local official, 22 March 2016.

39 Personal communication, local official, 22 March 2016.

Index